Dynamical Systems

An International Symposium

Volume 2

International Symposium on Dynamical Systems, Brown University, 1974.

Dynamical Systems

An International Symposium
Volume 2

Edited by

Lamberto Cesari   .

Department of Mathematics
University of Michigan
Ann Arbor, Michigan

Jack K. Hale

Joseph P. LaSalle

Lefschetz Center for Dynamical Systems
Division of Applied Mathematics
Brown University
Providence, Rhode Island

ACADEMIC PRESS   New York   San Francisco   London   1976
A Subsidiary of Harcourt Brace Jovanovich, Publishers

ACADEMIC PRESS, INC.
111 Fifth Avenue, New York, New York 10003

United Kingdom Edition published by
ACADEMIC PRESS, INC. (LONDON) LTD.
24/28 Oval Road, London NW1

Library of Congress Cataloging in Publication Data
Main entry under title:

Dynamical systems.

    Proceedings of an international symposium held at
Brown University, Aug. 12-16, 1974.
    Bibliography: p.
    Includes index.
    1.   Differential equations—Congresses.
2.   Topological dynamics—Congresses.   3.   Differentiable
dynamical systems—Congresses.   I.   Cesari, Lamberto.
II.   Hale, Jack K.   III.   LaSalle, Joseph P.
QA371.D9   515   75-13095
ISBN 0-12-164902-4   (v. 2)

Contents

Chapter 1   QUALITATIVE THEORY

CONTENTS

Chapter 2 GENERAL THEORY

Chapter 3 EVOLUTIONARY EQUATIONS

Chapter 4 FUNCTIONAL DIFFERENTIAL EQUATIONS

Chapter 5   TOPOLOGICAL DYNAMICAL SYSTEMS

Chapter 6 ORDINARY DIFFERENTIAL AND VOLTERRA EQUATIONS

List of Contributors

Numbers in parentheses indicate the pages on which the authors' contributions begin.

ZVI ARTSTEIN (73), Lefschetz Center for Dynamical Systems, Division of Applied Mathematics, Brown University, Providence, Rhode Island

PREM N. BAJAJ (191), Department of Mathematics, Wichita State University, Wichita, Kansas

JOHN M. BALL* (91), Lefschetz Center for Dynamical Systems, Division of Applied Mathematics, Brown University, Providence, Rhode Island

NAM P. BHATIA (187, 197), Division of Mathematics and Physics, University of Maryland, Baltimore County, Baltimore, Maryland

MARTIN BRAUN (37), Lefschetz Center for Dynamical Systems, Division of Applied Mathematics, Brown University, Providence, Rhode Island

ROGER W. BROCKETT (227), Division of Engineering and Applied Physics, Harvard University, Cambridge, Massachusetts

PIERRE CHARRIER (81), U.E.R. de Mathématiques et Informatique, Université de Bordeaux, Talence, France

RICHARD DATKO (95), Department of Mathematics, Georgetown University, Washington, D.C.

N. A. DE MOLFETTA (127), Universidade Federal de São Carlos, Sao Paulo, Brazil

RODNEY D. DRIVER (115), Department of Mathematics, University of Rhode Island, Kingston, Rhode Island

ROBERT W. EASTON (1), Department of Mathematics, University of Colorado, Boulder, Colorado

RICHARD H. ELDERKIN (7, 13), Department of Mathematics, Pomona College, Claremont, California

JOHN R. GRAEF (275), Department of Mathematics, Mississippi State University, Mississippi State, Mississippi

JAMES M. GREENBERG (153), Department of Mathematics, State University of New York at Buffalo, Amherst, New York

OKAN GUREL (255), International Business Machines Corporation, White Plains, New York

---

* Present address: Department of Mathematics, Heriot-Watt University, Edinburgh, Scotland.

JOHN R. HADDOCK (271), Department of Mathematical Sciences, Memphis State University, Memphis, Tennessee

C. S. HARTZMAN (19), Department of Mathematics, Dalhousie University, Halifax, Nova Scotia, Canada

TERRY L. HERDMAN* (307), Department of Mathematics, University of Oklahoma, Norman, Oklahoma

CHARLES HOLLAND (77), Department of Mathematics, Purdue University, West Lafayette, Indiana

R. R. HUILGOL† (57), Department of Mechanics and Mechanical Aerospace Engineering, Illinois Institute of Technology, Chicago, Illinois

CARLOS IMAZ (123), Centro de Investigacion y de Estudios Avanzados del Instituto Politécnico Nacional, México D.F., Mexico

A. F. IZE (127), Instituto de Ciências Matemáticas de São Carlos–U.S.P., Sao Paulo, Brazil

R. KANNAN (67), Department of Mathematics, Michigan State University, East Lansing, Michigan

JAMES L. KAPLAN (137), Department of Mathematics, Boston University, Boston, Massachusetts

F. KAPPEL (103), University of Würzburg, Würzburg, Germany

JUNJI KATO (217), Mathematical Institute, Tohoku University, Sendai, Japan

B. S. LALLI (265), Department of Mathematics, University of Saskatchewan, Saskatoon, Saskatchewan, Canada

JOSEPH P. LASALLE (291), Lefschetz Center for Dynamical Systems, Division of Applied Mathematics, Brown University, Providence, Rhode Island

ANDRZEJ LASOTA (47), Department of Mathematics, Jagellonian University, Krakow, Poland

M. J. LEITMAN (143), Department of Mathematics and Statistics, Case Western Reserve University, Cleveland, Ohio

TIEN-YIEN LI (203), Department of Mathematics, The University of Utah, Salt Lake City, Utah

JAMES C. LILLO (109), Division of Mathematical Sciences, Purdue University, West Lafayette, Indiana

N. G. LLOYD‡ (233), Department of Mathematics, St. John's College, Cambridge, England

---

* Present adress: Department of Mathematics, Virginia Polytechnic Institute and State University, Blacksburg, Virginia.

† Present address: School of Mathematical Sciences, The Flinders University of South Australia, Bedford Park, Australia.

‡ Present address: Department of Mathematics, University of Wales, Aberystwyth, U.K.

ORLANDO LOPES (*159*), Campinas, Sao Paulo, Brazil

CLEMENT A. W. MCCALLA (*163*), Department of Mathematics, Howard University, Washington, D.C.

T. MATSUMOTO (*285*), Department of Electrical Engineering, Waseda University, Shinjuku, Tokyo, Japan

V. J. MIZEL (*143*), Department of Mathematics, Carnegie-Mellon University, Pittsburgh, Pennsylvania

JAMES MURDOCK (*25*), Department of Mathematics, City College of New York, New York

DAVID R. NAUGLER (*29*), Department of Mathematics, Dalhousie University, Halifax, Nova Scotia, Canada

M. NISHIHAMA (*187, 197*), Division of Mathematics and Physics, University of Maryland, Baltimore County, Baltimore, Maryland

J. A. NOHEL (*297*), Department of Mathematics, University of Wisconsin, Madison, Wisconsin

ROGER D. NUSSBAUM (*99*), Department of Mathematics, Rutgers University, New Brunswick, New Jersey

CZESŁAW OLECH (*63*), Institute of Mathematics, Polish Academy of Sciences, Warsaw, Poland

CHONG-PIN ONG* (*41*), Department of Mathematics, University of California, Berkeley, California

E. N. ONWUCHEKWA† (*291*), Lefschetz Center for Dynamical Systems, Division of Applied Mathematics, Brown University, Providence, Rhode Island

R. H. PLAUT‡ (*279*), Lefschetz Center for Dynamical Systems, Division of Applied Mathematics, Brown University, Providence, Rhode Island

T. G. PROCTOR (*261*), Department of Mathematical Sciences, Clemson University, Clemson, South Carolina

R. S. RAMBALLY (*265*), Department of Mathematics, University of Saskatchewan, Saskatoon, Saskatchewan, Canada

COKE S. REED (*207*), Mathematics Department, Auburn University, Auburn, Alabama

R. REISSIG (*223*), Institut für Mathematik, Ruhr-Universität, Bochum, Germany

---

* Present address: Department of Mathematics, Iowa State University, Ames, Iowa.

† Present address: Ministry of Economic Development, Port Harcourt, Nigeria.

‡ Present address: Department of Civil Engineering, Virginia Polytechnic Institute and State University, Blacksburg, Virginia.

R. CLARK ROBINSON (33), Department of Mathematics, Northwestern University, Evanston, Illinois

HILDEBRANDO MUNHOZ RODRIGUES* (249), Instituto de Ciências Matemáticas de São Carlos, Universidade de Sao Paulo, Sao Paulo, Brazil

N. ROUCHE (133), Institut de Mathématique, University of Louvain, Louvain-la-Neuve, Belgium

ROBERT J. SACKER (175), Department of Mathematics, University of Southern California, Los Angeles, California

PABLO M. SALZBERG† (211), Instituto de Matemática, Universidad Católica de Chile, Santiago, Chile

PETER SEIBERT‡ (181), Instituto de Matemática, Universidad Católica de Chile, Santiago, Chile

GEORGE SEIFERT (303), Department of Mathematics, Iowa State University, Ames, Iowa

D. F. SHEA (297), Department of Mathematics, University of Wisconsin, Madison, Wisconsin

YASUTAKA SIBUYA (243), School of Mathematics, University of Minnesota, Minneapolis, Minnesota

PAUL W. SPIKES (275), Department of Mathematics, Mississippi State University, Mississippi State, Mississippi

C. C. TRAVIS (147), Mathematics Department, The University of Tennessee, Knoxville, Tennessee

TARO URA (169), Department of Mathematics, Faculty of Science, Kobe University, Kobe, Japan

MINORU URABE** (237), Department of Mathematics, Faculty of Science, Kyushu University, Fukuoka, Japan

J. A. WALKER (87), Department of Mechanical Engineering, Northwestern University, Evanston, Illinois, and Lefschetz Center for Dynamical Systems, Division of Applied Mathematics, Brown University, Providence, Rhode Island.

G. F. WEBB (147), Mathematics Department, Vanderbilt University, Nashville, Tennessee

---

* Present address: Division of Applied Mathematics, Brown University, Providence, Rhode Island.

† Present address: Departmento de Matemáticas, Universidad Simón Bolivar, Caracas, Venezuela.

‡ Present address: Departmento de Matemáticas, Universidad Simón Bolivar, Sartenejas, Baruta Edo Miranda, Venezuela.

** Deceased.

F. WESLEY WILSON, JR. (*13*), Department of Mathematics, University of Colorado, Boulder, Colorado

JAMES A. YORKE (*137, 203*), Institute for Fluid Dynamics and Applied Mathematics, University of Maryland, College Park, Maryland

TARO YOSHIZAWA (*217*), Mathematical Institute, Tohoku University, Sendai, Japan

Preface

The International Symposium on Dynamical Systems, of which this volume is the proceedings, was held at Brown University, August 12–16, 1974 and was the formal occasion for dedicating the Lefschetz Center for Dynamical Systems to the memory of Solomon Lefschetz. The central theme of the symposium was the manner in which the theory of dynamical systems continues to permeate current research in ordinary and functional differential equations, and how this approach and the techniques of ordinary differential equations have begun to influence in a significant way research on certain types of partial differential equations and evolutionary equations in general. This volume provides an exposition of recent advances, present status, and prospects for future research and applications.

The editors and the Lefschetz Center for Dynamical Systems wish to thank the Air Force Office of Scientific Research, the Army Research Office (Durham), the National Science Foundation, the Office of Naval Research, and Brown University for the generous support that made this symposium possible. The editors were responsible for the program, and we wish here to express on behalf of all the participants our appreciation to H. Thomas Banks, Ettore F. Infante, and Constantine Dafermos for their planning and organization of the meeting.

Contents of Volume 1

Chapter 1 : QUALITATIVE THEORY

Some Qualitative Aspects of the Three-Body Flow*

ROBERT W. EASTON
Department of Mathematics
University of Colorado, Boulder, Colorado

The purpose of this paper is to outline a way in which the recent theory of isolating blocks may be applied to the Newtonian planar three-body problem. My motivation in part is to test the strength of the theory. Of course, a full understanding of the three-body problem also requires application of the theory of quasi-periodic motion and perhaps a number of other as yet undiscovered ideas.

1. The Model

For $j = 1$, 2, 3 let q_j and p_j specify the position and momentum of a particle with mass $m_j > 0$ in the plane. Let $Q = (q_1, q_2, q_3) \in (R^2)^3$ and $P = (p_1, p_2, p_3) \in (R^2)^3$. Define

$$\Delta = \{Q \in R^6 : q_i = q_j \text{ for some } i \neq j\},$$
$$U(Q) = m_1 m_2 |q_1 - q_2|^{-1} + m_2 m_3 |q_2 - q_3|^{-1} + m_3 m_1 |q_3 - q_1|^{-1},$$
$$T(P) = \tfrac{1}{2}[m_1^{-1}|p_1|^2 + m_2^{-1}|p_2|^2 + m_3^{-1}|p_3|^2],$$
$$H(Q, P) = T(P) - U(Q).$$

Newton's model for the motion of three bodies in the plane under the influence of their mutual gravitational attraction consists of the manifold $(R^6 - \Delta) \times R^6$ of "states" of the system together with the equations of motion

$$(1.1) \qquad \dot{q}_k = m_k^{-1} p_k, \qquad \dot{p}_k = \sum_{j \neq k} m_j m_k |q_j - q_k|^{-3}(q_j - q_k).$$

* Partially supported by NSF grant GP-38585.

The equations of motion form a Hamiltonian system of differential equations with Hamiltonian function $H(Q, P)$ defined above.

It is well known that Newton's equations of motion conserve energy, momentum, and angular momentum. Without loss of generality we suppose that with respect to our coordinates (Q, P) the center of mass of the system is at the origin $(\sum_{k=1}^{3} m_k q_k = 0)$ and the total momentum is zero $(\sum_{k=1}^{3} p_k = 0)$.

(1.2) Definitions. For h, ω real numbers, define

$$J(\omega) = \left\{ (Q, P) : Q \neq 0, \ \sum_{k=1}^{3} m_k q_k = 0, \ \sum_{k=1}^{3} P_k = 0, \ \sum_{k=1}^{3} q_k \times p_k = \omega \right\},$$

$$M(h, \omega) = \{(Q, P) \in J(\omega) : H(Q, P) = h\}.$$

The conservation laws referred to above imply that solutions of (1.1) that start on a surface $M(h, \omega)$ remain on this surface for as long as they are defined. Solutions of (1.1) may, in fact, cease to exist after a finite time due to collisions of the bodies.

The problem of whether solutions can be analytically continued beyond collision was studied by Levi-Civita, Sundman, Siegel, and others. Recently a geometric definition of regularization was given [5] using isolating blocks. This definition was used in studying binary collisions of the planar three-body problem in [7] and in studying triple collisions in [11]. Space prevents a further discussion of these results.

A first step in the qualitative study of the three-body problem might be to describe the topology of the manifolds $M(h, \omega)$. This has been done for the planar problem [6, 17], and for the nonplanar problem [3, 9]. We want to concentrate now on describing the flow generated by solutions of (1.1) on the invariant manifolds $M(h, \omega)$.

The moment of inertia $I(Q) = \sum_{k=1}^{3} m_k |q_k|^2$ of the system satisfies the well-known Lagrange–Jacobi identity $\ddot{I}(Q) = U(Q) + 2h$. Thus for $h \geq 0$ all solutions that do not end in collision are unbounded. For this reason we restrict our attention to the flow on the manifolds $M(h, \omega)$ where $h < 0$. For each choice of $h < 0$, ω, and the three masses m_1, m_2, m_3, we have the problem of describing the flow on $M(h, \omega)$. We will also require $\omega \neq 0$ in order to avoid triple collisions of the bodies.

2. Isolating Blocks for the Three-Body Flow

A general approach to the study of a flow suggested by Charles Conley is as follows: First, locate certain isolated invariant sets of the flow; next, describe the phase portrait locally near these sets; next, establish how the invariant sets are tied together by heteroclinic and homoclinic orbits; and finally, describe the phase portrait near the union of the invariant sets and the heteroclinic and homoclinic orbits. This union may again be an isolated invariant set, in which case the process may be repeated. This is the approach we will apply to the three-body flow.

(2.1) Definitions. A (closed) invariant set K for a flow φ on a metric space X is *isolated* if there exists a neighborhood U of K in X such that K is the maximal invariant subset of U. An orbit is *heteroclinic* from an invariant K_1 to an invariant set K_2 if its alpha limit set is contained in K_1 and its omega limit set is contained in K_2. A *cycle* of heteroclinic orbits is a collection of invariant sets $K_0, \ldots, K_l$ with $K_0 = K_l$, together with a collection of orbits heteroclinic from K_j to K_{j+1} for $j = 0, \ldots, l-1$. An *isolating block* for the flow φ is roughly a closed subset $B \subset X$ such that the functions

$$\sigma^+(x) = \sup\{t \geq 0 : \varphi(s, x) \in B \text{ for } s \in [0, t]\},$$

$$\sigma^-(x) = \inf\{t \leq 0 : \varphi(s, x) \in B \text{ for } s \in [t, 0]\},$$

are continuous from B to $[0, \infty]$ and to $[-\infty, 0]$, respectively. For a further discussion of isolating blocks and their uses see [4].

(2.2) Definition. $B_0 = \{(Q, P) \in M(h, \omega) : I(Q) \leq \delta\}$. Choose δ sufficiently small so that $I(Q) \leq \delta$ implies that $U(Q) + 2h > 0$. Thus by the Lagrange–Jacobi identity we have $\ddot{I} > 0$ on B_0 and it follows that B_0 is an isolating block for the three-body flow on $M(h, \omega)$. For δ small, B_0 has three components each diffeomorphic to $(0, \delta] \times S^1 \times S^1 \times R^3$.

(2.3) Definition.

$$B_\infty(3) = \{(Q, P) \in M(h, \omega) : |q_3| \geq c, |q_1 - q_2| \leq |q_1 - q_3|,$$
$$|q_1 - q_2| \leq |q_2 - q_3|\}.$$

$B_\infty(3)$ requires configurations Q of the bodies, where the third body is far from the first two. Similarly define sets $B_\infty(1)$ and $B_\infty(2)$.

For C sufficiently large one can compute that $\ddot{\rho} < 0$ on $B_\infty(3)$, where $\rho(Q, P) = |q_3|$. It follows that $B_\infty(3)$ is an isolating block for the flow on $M(h, \omega)$. Similarly, $B_\infty(1)$ and $B_\infty(2)$ are isolating blocks for the flow and one can show that each of these blocks is diffeomorphic to $[c, \infty) \times S^1 \times S^1 \times R^3$. The flow on $B_\infty(3)$ can be analyzed using the two-body problem as a model. The first two bodies move in approximately elliptic orbits perturbed by the third body, which is far away. The third body is influenced by a force that is approximately the force of a fictitious body of mass $(m_1 + m_2)$ located at $(m_1 q_1 + m_2 q_2)/(m_1 + m_2)$. Hence the total motion is approximately the combined motion of two-body systems. It is important to characterize the set of orbits that enter $B_\infty(3)$ and for which $|q_3(t)|$ tends (monotonically) to infinity as $t \to \infty$. I believe these orbits form a manifold, which I will call $W^s(3, \infty)$, the "stable manifold" of infinity through $B_\infty(3)$. Similarly define $W^s(j, \infty)$ for $j = 1, 2$ and $W^u(j, \infty)$ for $j = 1, 2, 3$. Here $W^u(j, \infty)$ denotes the set of orbits for which $|q_j(t)|$ tends to infinity as $t \to -\infty$ through the block $B_\infty(j)$. The $W^u(j, \infty)$ form the "unstable manifold" of infinity.

For each $\omega \neq 0$ there are five periodic orbits of the three-body flow on the constant angular momentum surface $J(\omega)$. These periodic orbits were discovered by Euler and Lagrange. For two of these solutions the bodies form an equilateral triangle that rotates at a constant angular velocity in the plane. Let $h_0 = h_0(\omega)$ denote the energy of these solutions, which therefore are contained in $M(h_0, \omega)$. For the three other solutions, the bodies are situated along a line that spins at constant angular velocity. Let $h_j = h_j(\omega)$ denote the energy of the collinear periodic solution, where the jth mass is located between the other two.

One can show that for $\varepsilon > 0$ sufficiently small there exists a block $B(h_j, \varepsilon)$ for the flow on $M(h_j + \varepsilon, \omega)$ that lies near the jth collinear periodic solution for $j = 1, 2, 3$. $B(h_j, \varepsilon)$ is diffeomorphic to $S^1 \times [0, 1] \times S^4$. The maximal invariant subset K_j of $B(h_j, \varepsilon)$ is diffeomorphic to $S^1 \times S^3$ and the stable and unstable manifolds of this invariant set are diffeomorphic to $S^1 \times S^3 \times R^1$. These results follow from studying the linearized equations near the jth collinear periodic solution and by applying a perturbation argument.

It may be possible to apply the results of [10] here to do a fine analysis of the flow on these invariant sets.

3. Unfinished Work

I have not as yet made progress on analyzing how the stable and unstable manifolds of infinity intersect the stable and unstable manifolds of the invariant sets K_j near the collinear periodic solutions. References [1] and [16] are relevant here, and I conjecture that cycles of heteroclinic orbits do exist.

The study of the flow near a cycle will probably require some new ideas. Studies of this sort have been made in the case where the invariant sets involved were periodic orbits [2, 8, 14, 18].

Further questions involving the bifurcation of the phase portrait as the masses of the bodies change can be asked. Here the theory concerning continuation of invariant sets [12] may apply. In particular, the phase portrait of solutions near the equilateral triangle solution probably changes at the mass ratio

$$(m_1 m_2 + m_2 m_3 + m_3 m_1)/(m_1 + m_2 + m_3)^2 = 1/27$$

(see Section 18 of [15]).

REFERENCES

[1] Alekseev, V. M., On the capture orbits for the three-body problem for negative energy constant, *Uspehi Mat. Nauk* **24** (1969), 185–186.

[2] Birkhoff, G. D., Nouvelles recherches sur les systèmes dynamiques, *Mem. Pont. Acad. Sci. Novi Lyncaei* **1** (1935), 85–216.

[3] Cabral, H., On the integral manifolds of the *n*-body problem, *Invent. Math.* **20** (1973), 59–72.

[4] Conley, C. C., and R. W. Easton. Isolated invariant sets and isolating blocks, *Trans. Amer. Math. Soc.* **158** (1971), 35–61.

[5] Easton, R. W., Regularization of vector fields by surgery, *J. Differential Equations* **10** (1971).

[6] Easton, R. W., Some topology of the 3-body problem, *J. Differential Equations* **10** (1971).

[7] Easton, R. W., Topology of the regularized integral surfaces in the 3-body problem, *J. Differential Equations* **12** (1972).

[8] Easton, R. W., Isolating blocks and symbolic dynamics, *J. Differential Equations* **17** (1975).

[9] Easton, R. W., Some topology of *n*-body problems, *J. Differential Equations* **19** (1975).

[10] Graff, S., On the conservation of hyperbolic invariant tori for Hamiltonian systems, *J. Differential Equations* **15** (1974).

[11] McGehee, R., Triple collision in the colinear 3-body problem, *Invent. Math.* (1975).

[12] Montgomery, J., Cohomology of isolated invariant sets under perturbation, *J. Differential Equations* **13** (1973).

[13] Moser, J., "Stable and Random Motions in Dynamical Systems" (Ann. Math. Studies, No. 77). Princeton Univ. Press, Princeton, New Jersey, 1973.

[14] Rod, D., Pathology of invariant sets in the monkey saddle, *J. Differential Equations* **14** (1973).

[15] Siegel, C. L., and J. Moser. "Lectures on Celestial Mechanics." Springer-Verlag, Berlin and New York, 1971.

[16] Sitnikov, K., Existence of oscillating motions for the three-body problem, *Dokl. Akad. Nauk SSSR* **133** (1960).

[17] Smale, S., Topology and mechanics I and II, *Invent. Math.* **10, 11** (1970).

[18] Smale, S., "Diffeomorphisms with Many Periodic Points, Differential and Combinatorial Topology." Princeton Univ. Press, Princeton, New Jersey, 1965.

Separatrix Structure for Regions
Attracted to Solitary Periodic Solutions

RICHARD H. ELDERKIN*
Department of Mathematics
Pomona College, Claremont, California

This paper is concerned with separatrix structure near solitary periodic solutions in 3-space. The spirit is similar to [1]. Under strict hypotheses, regions attracted to the periodic solution are identified by boundary type. Of four possible types, three have no internal separatrices. Separatrix structure in the remaining type is analyzed. For purposes of comparison, a weaker hypothesis is considered and examples are given to point out differences between the weaker and strict hypotheses.

Let $\phi(x, t)$ be the solution flow of a smooth vector field $\dot{x} = f(x)$ on $\mathbf{R}^3$. It will be convenient to have f extended to the 3-sphere S^3, which we view as $\mathbf{R}^3 \cup \{\infty\}$, with $f(\infty) = 0$. The α- and ω-limits, given respectively by

$$\alpha(x) = \{y \,|\, \exists t_n \to \infty \text{ with } \phi(x, -t_n) \to y\}$$
$$\omega(x) = \{y \,|\, \exists t_n \to \infty \text{ with } \phi(x, t_n) \to y\}$$

will be viewed as functions,

$$\alpha, \omega: S^3 \to F(S^3)$$

where $F(S^3)$ is the topological space of closed nonempty subsets of S^3 with the Hausdorff metric topology.

Let

$$\mathscr{P}(\phi) = \{x \in S^3 \,|\, \alpha, \omega \text{ are continuous at } x \text{ and } \exists \text{ nbhd } N \text{ of } x$$
$$\text{satisfying } \forall \varepsilon > 0 \; \exists T > 0 \text{ such that } y \in N \text{ and } t > T$$
$$\Rightarrow d(\phi(y, -t), \alpha(x)) < \varepsilon \text{ and } d(\phi(y, t), \omega(x)) < \varepsilon\}$$

(set of parallel flow)

* This research was partially supported by the U.S. Army Research Office (Durham) under grants DA-ARO-D-31-124-71-G12-S2 and DA-ARO-D-31-124-73-G130 while the author was at Brown University, Providence, Rhode Island.

7

A separatrix is a trajectory in the set $\mathcal{S}(\phi) = S^3 - \mathcal{P}(\phi)$. Because of examples where $\mathcal{S}(\phi) = S^3$, it is convenient to partition $\mathcal{S} = \mathcal{S}(\phi)$:

$$\mathcal{S} = \mathcal{S}_0 \cup \mathcal{S}_1 \cup \mathcal{S}_2$$

where

$$\mathcal{S}_0 = \mathcal{S}_0(\phi) = \{x \,|\, \alpha \text{ or } \omega \text{ is discontinuous at } x\}$$
$$\mathcal{S}_1 = \mathcal{S}_1(\phi) = \{x \in \mathcal{S} - \mathcal{S}_0 \,|\, \exists x_n \to x, \, t_n \to \infty \text{ with}$$
$$\lim \phi(x_n, -t_n) \notin \alpha(x) \text{ or } \lim \phi(x_n, t_n) \notin \omega(x)\}$$
$$\mathcal{S}_2 = \mathcal{S}_2(\phi) = \mathcal{S} - (\mathcal{S}_0 \cup \mathcal{S}_1)$$

These sets are clearly ϕ-invariant. Furthermore, a close examination of the definition reveals that $\mathcal{S}_2$ is contained in the closure of $\mathcal{S}_0 \cup \mathcal{S}_1$.

We suppose that ϕ has a nontrivial periodic solution γ, homeomorphic to the circle S^1. We furthermore suppose that γ is solitary (cf. [2]), that is, there exists a compact neighborhood U of γ (neighborhood of solitude) satisfying (i) $\phi(x, (-\infty, 0]) \subset U$ implies $\alpha(x) = \gamma$, and (ii) $\phi(x, [0, \infty)) \subset U$ implies $\omega(x) = \gamma$.

Lemma. Let γ be a solitary periodic solution of ϕ. Then a neighborhood of solitude U may be chosen to satisfy:

(i) U is diffeomorphic with $D^2 \times S^1$ such that $\gamma = \{0\} \times S^1$, where D^2 is the closed unit disk in the plane;

(ii)   There is a Riemannian metric on U such that when the vector field f is represented in $D^2 \times S^1$ coordinates as $f(z, \theta) = (\dot{z}, \dot{\theta})$, then $\dot{\theta} \equiv 1$;

(iii) f has generic contact with ∂U (∂ = boundary of a manifold).

In this case, U is called a generic neighborhood of solitude.

The vector field f has *generic contact* with the submanifold ∂U in $\mathbf{R}^3$ if the following conditions are satisfied: the subset of ∂U where f if tangent to ∂U is either empty or a finite set of circles collectively denoted by τ. Each circle separates a region of egress from a region of ingress of f relative to U, with the component of f normal to ∂U (it is 0 precisely on τ) having nonzero derivative in a direction in ∂U transverse to τ, everywhere on τ. Furthermore, the subset of τ where f is tangent to τ is either empty or a finite set of points, collectively denoted χ (these are the crossing tangencies). Each point of χ separates an open subarc of τ where f points toward a region of ingress from an open subarc of τ where f points toward a region of egress. Finally, the component of f tangent to τ (is 0 precisely on χ and) has nonzero derivative in the direction of τ, everywhere on χ.

The proof of the lemma is fairly straightforward, except for part (iii), which can be derived from a theorem of Percell [3, Theorem 2.5].

Let us fix a neighborhood of solitude U satisfying the lemma's conclusion. Then we may partition U as follows:

$$U_\pm = \{x \in U \mid \phi(x, \pm t)U \text{ for all } t \geq 0\}$$
$$E = U_+ \cap U_- - \gamma \qquad \text{(elliptic set)}$$
$$A_+ = U_+ - (E \cup \gamma) \qquad \text{(positively attracted set)}$$
$$A_- = U_- - (E \cup \gamma) \qquad \text{(negatively attracted set)}$$
$$H = U - (U_+ \cup U_-) \qquad \text{(hyperbolic set)}$$

On the local analysis we are developing, we should like the boundaries of E, H, $A_\pm$ to be composed of separatrices. However, due to the global nature of the definition of separatrix, this is usually not so, since E, H, $A_\pm$ are relative to our choice of U. Hence, a modification of definition is necessary for the local situation. Our naïve viewpoint is that ∂U should be like a point at infinity, and any trajectory that leaves U is lost from our view. Therefore, let g be a smooth nonnegative-valued function on $\mathbf{R}^3$ with $g^{-1}(0) = \mathbf{R}^3 - \text{Int}(U)$ (Int = topological interior), and let ψ be the solution flow to the vector field $g \cdot f$. Use σ to denote generically a maximum trajectory segment of ϕ $[\phi(x, I)$, where I is the maximum closed, not necessarily bounded interval such that $\phi(x, I) \subset U]$, and note that each σ is a union of ψ-trajectories. Say σ is a separatrix of ϕ relative to U if it contains a ψ-separatrix. Our notation is

$$\mathscr{S}(\phi, U) = \{x \in \sigma \mid \sigma \cap \mathscr{S}(\psi) \neq \phi\}$$
$$\mathscr{S}_0(\phi, U) = \{x \in \sigma \mid \sigma \cap \mathscr{S}_0(\psi) \neq \phi\}$$
$$\mathscr{S}_1(\phi, U) = \{x \in \sigma \mid \sigma \not\subset \mathscr{S}_0(\phi, U) \text{ and } \sigma \cap \mathscr{S}_1(\psi) \neq \phi\}$$
$$\mathscr{S}_2(\phi, U) = \mathscr{S}(\phi, U) - (\mathscr{S}_0(\phi, U) \cup \mathscr{S}_1(\phi, U))$$
$$\mathscr{P}(\phi, U) = U - \mathscr{S}(\phi, U)$$

Henceforth in this chapter, we will be concerned only with separatrices of ϕ relative to U. Therefore, we write $\mathscr{S}$, $\mathscr{S}_0$, $\mathscr{S}_1$, $\mathscr{S}_2$, and $\mathscr{P}$ to denote the respective sets above.

A point $x \in \partial U$ is an internal tangency for the trajectory segment in U through it if $\phi(x, (0, \delta))$ and $\phi(x, (-\delta, 0))$ are contained in $\text{Int}(U)$ for some $\delta > 0$.

Lemma. Any trajectory segment in $U_\pm$ with an internal tangency is contained in $\mathscr{S}_0$.

The proof follows easily from generic contact of f with ∂U.

Lemma. Let $x \in b(A_+)$ (b = topological boundary in U). Then either
(a) $x \in E$ or (b) $x \in b(H)$. Let $L = \sup\{t \mid \phi(x, [0, t]) \subset b(A_+)\}$. In case (a),
L is infinite. In case (b), L may be finite or infinite. In the finite case,
$\phi(x, L)$ is an internal tangency with $\phi(x, [0, L]) \subset b(H)$ and $\phi(x, (L, \infty))$
$\subset \text{Int}(A_+)$. In the infinite case, $\phi(x, \mathbf{R}^+) \subset b(A_+) \cap b(H)$ and there is an
internal tangency of the form $\phi(x, T)$ such that $\phi(x, (T, \infty))$ contains no
internal tangencies and such that if $x_n \to \phi(x, t)$ with $t > T$ and $x_n \in H$
(all n), then $\phi(x_n, [0, t_n)) \subset H$ for $t_n \to +\infty$.

For many periodic solutions, there are choices of solitude so that $b(A_+)$ is
positively flow-invariant, as well as choices so that this is not the case.
The latter situations are beyond our present scope, so we now assume:

A. $\phi(b(A_+), \mathbf{R}^+) \subset b(A_+)$.

We also assume the following strict manifold hypothesis:

H. $\mathscr{S}_0 \cup \mathscr{S}_1 - \gamma$ is an embedded submanifold of U.

By "submanifold of U" we mean that for each point x of $\mathscr{S}_0 \cup \mathscr{S}_1 - \gamma$
either (a) $x \in \text{Int}(U)$ and the usual condition holds, or (b) $x \in \partial U$ and either
the usual condition holds in U, or else there is a submanifold chart in
the double of U for $\mathscr{S}_0 \cup \mathscr{S}_1 - \gamma$ at x. This weakened definition allows
$\mathscr{S}_0$ or $\mathscr{S}_1$ to be internally tangent to ∂U.

For the main theorem we need some notation. Let $\Theta: U = D^2 \times S^1 \to S^1$
be the projection. Let A be a connected component of A_+; fix x in A;
let $\tilde{A}_t$ denote the component of $A \cap \text{Int}(D^2 \times \{\Theta(\phi(x, t))\})$ that contains
$\phi(x, t)$; and finally let $A_t = \text{Cl}(\tilde{A}_t) - \gamma$ (Cl = closure).

Main theorem. There is a $T \geq 0$ such that one of the following holds.

(a)   There is a homeomorphism $F: \bigcup_{t \geq T} A_t \to D^2 \times [0, \infty)$ such that
$F(A_t) = D^2 \times \{t - T\}$. In this case, $A_t \to \gamma$ as $t \to \infty$ in the Hausdorff metric
topology, and $\text{Int}(A) \subset \mathscr{P}$.

(b)   There is a homeomorphism $F: A \to (0, 1] \times S^1$. Here $F(A_t) \subset (0, 1]$
$\times \{e^{\pi i t/m}\}$, where m is a positive integer such that $\phi(A_t, mn) \subset A_t$ for every
positive integer n. F extends to a homeomorphism of $A \cup \gamma$ onto a
quotient space $[0, 1] \times S^1/R$, where R is the relation $(s_1, e^{t_1 \pi i})R(s_2, e^{t_2 \pi i})$ if
$s_1 = s_2 = 0$ and $t_2 = t_1 + n/m$ for some integer n and with m as before. γ is
mapped onto $\{0\} \times S^1/R$. Finally, $A \subset \mathscr{S}_0$.

(c)   There is a sequence (possibly void or finite) of nonoverlapping sub-intervals $\{(a_i, b_i)\}$ of $(0, 1)$ and continuous functions $C_i: (a_i, b_i) \to (-\infty, 0)$ with $C_i(u) \to -\infty$ as either $u \to a_i$ or $u \to b_i$ such that for each $t \geq T$, there is a homeomorphism $f_t: A_t \to \{(u, v) \in \mathbf{R}^2 \mid 0 \leq u \leq 1, \ v \leq 0, \text{ and } v \geq C_i(u)$ whenever $a_i < u < b_i\}$. Here $f_t(\partial U \cap A_t) = \{(u, v) \mid v = 0, \ 0 \leq u \leq 1\}$ and if $x_n \to \gamma$ in A_t and $(u_n, v_n) = f_t(x_n)$, then $v_n \to -\infty$. For each i, $0 \leq i \leq k$, the set $E_i = (\bigcup_{t \geq 0} f_t^{-1}(\text{graph}(C_i))) \cup \gamma$ is a tube, wrapping around γ and bounding a component of the elliptic set. Precisely E_i is the quotient space $S^1 \times S^1/R$, where R is the relation determined by a preferred point $w_0 \in S^1$ and a positive integer m with $(\alpha_1, \beta_1) R(\alpha_2, \beta_2)$ if $\alpha_1 = \alpha_2 = w_0$ and $\beta_2 = \beta_1 e^{\pi i n/m}$ for some positive integer n. The integer n also satisfies $\phi(A_t, mn) \subset A_t$ for positive integers n. However, A is not necessarily homeomorphic to $A_t \times S^1$.

Corollary. $\mathscr{S}_1 \cap A \neq \phi$ only if A is of type (c).

Theorem.   If C is a component of $A \cap \mathscr{S}_1$, then C has the same description as A in (b) for some possibly different m, and $\{y \mid \exists x \in C$ and $x_n \to x, \ t_n \to \infty$ such that $\phi(x_n, t_n) \to y\} = \gamma \cup (\bigcup E_i)$, where the union of E_i is such that if (a_j, b_j), (a_k, b_k), (a_l, b_l) are intervals of domain in C such that $b_j < a_k < a_l$ and such that (a_j, b_j) and (a_l, b_l) are involved in the above union, then (a_k, b_k) is also involved.

For contrast with the above theorems, it is instructive to consider the following weaker hypothesis: Assume $\mathscr{S}_0 \cup \mathscr{S}_1 - \gamma$ an immersed submanifold of U, such that each component of $\mathscr{S}_0 \cap (D^2 \times \{\theta\}) - \gamma$ or $\mathscr{S}_1 \cap (D^2 \times \{\theta\}) - \gamma$ is an embedded submanifold. A relevant example is constructed by suspending the composition of two diffeomorphisms, the first of which maps the region X_i onto X_{i+1} (Fig. 1) with no lateral

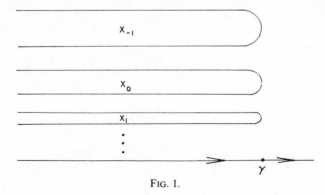

FIG. 1.

effect, and the second of which moves every point strictly to its right, excepting γ and the right endpoints of the X_i, which shall be fixed, and having only enough vertical displacement so that the X_i are invariant sets. Then $A_+ = \text{Cl}(\bigcup_i X_i)$, and the remainder is in H, save the ray emanating from γ to the right, which is in A_-.

REFERENCES

[1] R. Elderkin, Separatrix structure for elliptic flows, *Amer. J. Math.* **97** (1975), 221–247.
[2] R. Elderkin and F. W. Wilson, Jr., Solitary invariant sets, this volume.
[3] P. B. Percell, Structural stability on manifolds with boundary, *Topology* **12** (1973), 123–144.

Solitary Invariant Sets

RICHARD H. ELDERKIN
Department of Mathematics
Pomona College, Claremont, California

F. WESLEY WILSON, JR.
Department of Mathematics
University of Colorado, Boulder, Colorado

Let M denote a smooth (C^∞) n-manifold, let $f\colon M \times \mathbf{R} \to M$ be a C^r flow on M $(1 \le r \le \infty)$, and let R be a compact region in M that is a submanifold with boundary. Let I denote a closed invariant set of f that is interior to R. We have the following partition of the orbit segments in R:

$$R^\pm = \{x \in R \mid f(x, \pm t) \in R \text{ for all } t \ge 0\}$$
$$E = (R^+ \cap R^-) - I \qquad \text{(elliptic set)}$$
$$A^\pm = R^\pm - (E \cup I) \qquad \text{[positively (negatively, resp.) attracted set]}$$
$$H = R - (R^+ \cup R^-) \qquad \text{(hyperbolic set)}$$

Recent efforts to study this structure include the study of isolated invariant sets by Conley *et al.* [1–4, 6, 10] in which it is assumed that there are no elliptic orbits, and a paper by Percell [7] in which it is assumed that I is empty and every orbit is hyperbolic.

The authors are interested in finding a setting in which elliptic behavior can be studied. Examples of studies of very specialized elliptic behavior can be found in [5] and [9]. So far we have not required that I contain all of the recurrent behavior in R, and in general we shall prefer that there be no recurrence in the elliptic set. The condition that we impose on (R, I) is slightly stronger and requires that all limiting behavior in R is associated with I.

Definition 1. R is a *neighborhood of solitude* for I if every semiorbit of f that remains in R has its limit set contained in I. If I is an invariant set that has a neighborhood of solitude, then I is called a *solitary invariant set*.

Any isolated invariant set is solitary, and any isolating neighborhood is a neighborhood of solitude. In this case the elliptic set is always empty.

Also, any rest point of a planar flow, which is not a center, and which is isolated as a rest point, is a solitary invariant set, and many of these examples exhibit elliptic behavior.

In the study of isolated invariant sets, it is desirable to limit one's attention to certain specially structured isolating neighborhoods, the isolating blocks [3, 10]. We do not know what is the simplest structure for a neighborhood of solitude in general, but we have discovered certain simplifications, which are always possible, and these are the focus of our current attention.

Since R is a submanifold and f is a C^r flow, it is possible to discuss the tangencies of orbits with ∂R. Our first simplification of the structure of a neighborhood of solitude involves reducing the set along which ∂R is tangent to the flow. For this we use the notion of generic contact, which was introduced by Pugh in his study of the Poincaré index formula [8]. Before introducing the technical details, we pause to consider the example of a C^r isolating block [10]. Here the tangency set is either empty or a submanifold that separates the ingress region from the egress region in ∂R. At each point of the tangency set, the orbit is directed transverse to the tangency set and into the egress region. In general, it is not possible to choose the isolating block so that the tangency set is empty. For general neighborhoods of solitude, it is not only necessary to have a nonempty tangency set, but it may even be necessary that some orbits be tangent to this tangency set. Generic contact means that such tangencies of all orders have been minimized locally.

Definition 2. ∂R has *generic contact* with the flow f if there is a sequence of submanifolds $\partial R = M_1 \supset M_2 \supset \cdots \supset M_n$ such that for each $k = 1, \ldots, n$,

1. $M_k = \varnothing$ or codim $M_k = k$;
2. M_k is a C^{n-k} submanifold of M;
3. M_k is the set along which f is tangent to M_{k-1} and this tangency is generic, i.e.,

$$TM_k \pitchfork \text{Im}(Tf \,|\, TM_k) \quad \text{in} \quad TM_{k-1}|M_k$$

where T denotes the tangent functor and $\pitchfork$ denotes transversal intersection.

In particular, if we let $M_1{}^+$ and $M_1{}^-$ denote the respective ingress and egress regions of $\partial R = M_1$, then M_2 is the submanifold of M_1 that separates these regions. More generally, if we let $M_k{}^{\pm}$ denote the subsets of M_k along which the orbits are directed into $M_{k-1}^{\pm}$, then M_{k+1} is the submanifold of M_k that separates these regions. Pugh showed that it is always possible to achieve generic contact by small perturbations of the flow, which was adequate for his studies of the index. For general qualitative purposes, it is desirable to obtain generic contact by perturbing the embedding of ∂R in M. The authors have proved such a theorem by using the transversality approach of Abraham and the representation of mappings $f \to Tf$ on the set of C^r embeddings A_0^r of ∂R into M whose images separate M and are disjoint from the set of rest points of f. At the completion of this work, the authors became aware of the fact that this theorem had been proved (but not stated) by Percell using the methods of singularities of differentiable mappings [7; Proof of Theorem 2.5]. While it is not immediately clear that these approaches are equivalent, the authors have been able to establish this equivalence, and Percell's approach gives a much shorter proof.

Theorem 1. For $r > n$, the subset of A_0^r consisting of those embeddings, whose image has generic contact with f, is open and dense in A_0^r.

This theorem can be viewed as a general existence theorem for generalized cross sections for flows. In our case, it says that we can assume that all neighborhoods of solitude have generic contact with the flow. We note that if we had started with an isolated invariant set, just requiring that ∂R have generic contact with the flow would not be enough to ensure that we have an isolating block, since it would not remove internal tangencies in general. Certainly, the criteria for "simplest" neighborhood of solitude should be such that if the invariant set happened to be isolated, then the simplest neighborhood of solitude would be an isolating block.

Before introducing the second generic simplification of the choice of R, we wish to give an example of a neighborhood of solitude that is too complicated and a procedure for simplifying it. Consider the case where I is an asymptotically stable periodic solution of an autonomous ordinary differential equation in 3-space, and where R is a neighborhood of I that is bounded by some level surface of a Lyapunov function, i.e., $M_1^+ = R$, $M_1^- = \varnothing = M_2$. We construct a new R by stretching ∂R so that a rather tall, thin hump is formed in a direction normal to the old boundary. For this new R, M_1^- and M_2 are no longer empty. In fact, M_2 will have a single component that is a topological circle and reaches from the top of the hump to the base of the hump, and M_1^- is a topological disk that lies on the hump and interior to M_2. Since the flow is directed from M_2 into M_1^- near the top of the hump, and from M_2 into M_1^+ near the bottom of the hump, it follows that $M_3 \neq \varnothing$. In the simplest form of this example, $M_2{}^+$ and $M_2{}^-$ are topological arcs that are separated by two points that are all of $M_3 = M_3{}^+$. Thus we have a neighborhood of solitude for I that is more complicated than is necessary. In this case we can constructively remove the extra complications, and the procedure we shall use is interesting because it has wider application and because it leads us to the next way that a neighborhood of solitude can be generically simplified.

Observe that the orbits through the points of $M_2{}^+$ are internally tangent to M_1, and that the orbits through the points of M_3 are tangent to M_1 and cross from the exterior of R to the interior of R. Thus the orbits that originate in $M_2{}^+$ near the endpoints must also escape from R in small negative time. Thus we have a mapping defined from some points in $M_2{}^+$ to $M_1{}^+$ by associating each point to the point where its orbit first intersects M_1 in negative time.

Lemma. If the mapping described above is defined on an entire component A of $M_2{}^+$, then there is another neighborhood of solitude for I such that $M_2 = \varnothing$.

Proof. The image of this mapping is an arc A^* in $M_1{}^+$, whose endpoints are also the endpoints of A. Thus $A \cup A^*$ bounds a disk D in M_1 that contains M_2 in its interior. Let D^* denote the disk interior to R that is the union of the orbit segments from A to A^*. Then $D \cup D^*$ bounds a 3-disk U in R, and $R^* = R - U$ is a piecewise smooth neighborhood of solitude for I. Since the only lack of smoothness occurs near D^*, and since the orbits near D^* all enter the interior of R^*, it is possible to use Lyapunov function techniques to obtain a smooth neighborhood of solitude for I that has $M_2 = \varnothing$.

We note that this construction can be used to remove any bounding component of M_2 such that the indicated mapping is defined on some full component of $M_2{}^+$. This kind of construction can be used to remove some components of M_3 also. It can be attempted whenever there is a component A of M_2 that is bounded by points from $M_3{}^+$. The first point of crossing map extends to all of A unless one of the following situations occurs:

   (i) $A \cap R^- \neq \varnothing$;

   (ii) There is a point of A whose orbit is again internally tangent to M_1.

In the first case it is not likely that any kind of cancellation is possible.

The second case may or may not be an essential problem, and must be checked carefully.

The reason we have introduced this example is to illustrate that orbits internally tangent to M_1 at more than one point pose special problems. It turns out that Percell has also proved that we can generically minimize this behavior [7; Theorem 6.4]. We shall give a brief description of his approach, before stating the theorem in our context. Tangencies to M_1 occur at points of M_2. Consider the partition $N_k = M_k - M_{k+1}$ of M_2. If C is a component of some N_k, then the flow gives an immersion of $C \times \mathbf{R}$ into M. The multiply tangent orbits of the flow correspond to the nontrivial self-intersections of the images of this family of immersions. The precise setting is somewhat technical, but Percell uses essentially a transversality argument to show that, for most choices of R there will be a minimal structure of multiply tangent orbits. For example, it follows from dimensional arguments that in 3-space, there may be essential multiply tangent orbits, but that these will be isolated and at worst doubly internally tangent.

Definition 3. A *generic neighborhood of solitude R* for a solitary invariant set I is a neighborhood of solitude such that R has generic contact with the flow, and such that the structure of orbits that are multiply tangent with R is generically minimal.

Theorem 2. If f is a C^{n+1} flow on an n-manifold M, and if I is a solitary invariant set for f, then I has generic neighborhoods of solitude.

REFERENCES

[1] R. C. Churchill, Invariant sets which carry cohomology, *J. Differential Equations* **13** (1973), 523–550.

[2] R. C. Churchill, G. Pecelli, and D. L. Rod, Isolated unstable periodic orbits, *J. Differential Equations* **17** (1975), 329–398.

[3] C. Conley and R. Easton, Isolated invariant sets and isolating blocks, *Trans. Amer. Math. Soc.* **158** (1971), 35–61.

[4] R. Easton, Isolating blocks and symbolic dynamics, *J. Differential Equations* **17** (1975), 96–118.

[5] R. Elderkin, Separatrix structure for elliptic flows, *Amer. J. Math.* **97** (1975), 221–247.

[6] J. T. Montgomery, Cohomology of isolated invariant sets under perturbation, *J. Differential Equations* **13** (1973), 257–299.

[7] P. B. Percell, Structural stability on manifolds with boundary, *Topology* **12** (1973), 123–144.

[8] C. Pugh, A generalized Poincaré index formula, *Topology* **7** (1968), 217–226.

[9] F. W. Wilson, Elliptic flows are trajectory equivalent, *Amer. J. Math.* **92** (1970), 779–792.

[10] F. W. Wilson and J. A. Yorke, Lyapunov functions and isolating blocks, *J. Differential Equations* **13** (1973), 106–123.

Singular Points and Separatrices*

C. S. HARTZMAN†
Department of Mathematics
Dalhousie University, Halifax, Nova Scotia, Canada

0. Introduction

Bendixson in 1901 essentially completed the geometric classification of isolated singular points of a differential equation in the plane by recognizing a way in which an isolating disk neighborhood of any such singular point could be broken up into a finite number of sectors of these types—hyperbolic, elliptic, and parabolic (see [2] for a cogent explanation). Subsequent attempts at an analogous procedure for isolated singular points of differential equations in higher dimensions have met with little success. The purpose of this paper is to take a small step in that direction, preserving the spirit of Bendixson. To this end, in view of the fact that one would no doubt think of the boundaries of the sectors in the two-dimensional case as separatrices, we have formulated a definition of separatrix. The components of the complement of the set of separatrices are classified according to their topological behavior and this classification is further examined in relation to singular points. It is further shown that the set of separatrices can be decomposed into a disjoint union of manifolds.

The definition of separatrix offered here is done so in recognition of the fact that there are other definitions recently formulated, notably those of Markus [3] and Bhatia and Franklin [1]. No definition has yet gained currency. Bhatia and Franklin point out some of the shortcomings of Markus's definition.

Throughout the paper n-dimensional manifolds M^n will be differentiable of class C^r and flows $\alpha = \alpha(m_1 t): M^n \times \mathbb{R} \to M^n$ of class C^k, $k \geq r$. Given such a flow α on a manifold M^n, the following will be taken as the definition of transverse submanifold.

0.1. Definition. An embedded submanifold $M^{n-1} \subset M^n$ will be said to

---

* Complete proofs of the theorems in this paper will appear elsewhere.

† The author wishes to acknowledge the support of NRC Grant #A 8050.

be transverse to α if and only if there is a neighborhood U of M^{n-1} in M^n that is homeomorphic to $M^{n-1} \times (-\varepsilon, \varepsilon)$ in such a way as to map M^{n-1} onto $M^{n-1} \times \{0\}$ and map $\{\alpha(m_1 t): -\varepsilon < t < \varepsilon\}$ onto $m \times (-\varepsilon, \varepsilon)$ for all $m \in M^{n-1}$.

0.2. Definition. Two flows α on M^n and β on N^n will be called topologically equivalent if and only if there is a homeomorphism ϕ: $M^n \to N^n$ carrying each sensed (but not parametrized) orbit of α onto a sensed orbit of β.

The paper is intended as a summary of results, hence the word proof is to be read as indication of proof. Many of the facts will be stated without proof.

1. Definition of Separatrix

Given a flow α on a manifold M^n.

1.1. Definition. A subset P of M^n will be called a maximal parallel region if and only if

(i) There is a connected embedded submanifold $M^{n-1} \subset M^n$ transverse to α such that P is the smallest invariant subset of M^n containing M^{n-1}; or

(i') P is a union of an increasing sequence of regions $\{P_\alpha\}$ each topologically equivalent to some $M_\alpha^{n-1} \times S^1$, where M_α is an embedded connected submanifold of M^n; and

(ii) P is not properly contained in a region P_1 that is topologically equivalent to $M_1^{n-1} \times \mathbb{R}$ or $M_1^{n-1} \times S^1$ for some embedded connected submanifold M_1^{n-1} of M^n, where the orbits of $M_1^{n-1} \times \mathbb{R}$ or $M_1^{n-1} \times S^1$ are taken to be the usual straight-line or closed orbits.

Regions satisfying (i) will be called parallel regions.

1.2. Definition. A point $x \in M^n$ will be called separating if and only if

(i) $x \in \mathrm{cl}(\bigcup \bar{P} | P)$, the union being taken over all maximal parallel regions; and

(ii) x is not an interior point relative to M^n in the set of points lying in regions topologically equivalent to $M^{n-1} \times \mathbb{R}$ (or $M^{n-1} \times S^1$) for any embedded M^{n-1} unless $x \in \bar{P} \backslash P$ implies P is topologically equivalent to $M_1^{n-1} \times \mathbb{R}$ (or $M_1^{n-1} \times S^1$) for some embedded M_1^{n-1}.

Since maximal parallel regions are invariant and hence so are their closures, we have

1.3. Definition. The set of points $\{\alpha(p_1 t): t \in \mathbb{R}\}$ is called a separatrix if and only if p is separating. **S** will denote either the set of separating points or the set of separatrices.

1.4. Lemma. Every parallel region is contained in a maximal parallel region.

Proof. Let P be a parallel region that is not maximal. Then it is contained in a region P_1 topologically equivalent to $M_1^{n-1} \times \mathbb{R}$ (or $M_1^{n-1} \times S^1$). P_1 is a parallel region. Suppose P_1 is not maximal; then it is contained in a region P_2 topologically equivalent to $M_2^{n-1} \times \mathbb{R}$ (or $M_2^{n-1} \times S^1$). Repeat the procedure. $P \subset P_1 \subset P_2 \subset \cdots$. Let P_∞ be a largest region containing P that can be covered in this fashion. Claim, P_∞ is maximal. Clearly, (ii) of 1.1 is satisfied; hence all we must do is show that P_∞ is parallel. If P_∞ is a union of regions each topologically equivalent to $M_i^{n-1} \times S^1$, then P_∞ is maximal parallel.

We show P_∞ is parallel in case P_1 topologically equivalent to $M_1^{n-1} \times \mathbb{R}$. Then P_2 is topologically equivalent to $M_2^{n-1} \times \mathbb{R}$. Let $\phi_1(m) = \alpha(m, t(m))$ for all $m \in M_1^{n-1}$, where $t(m)$ is the time that $\alpha(m, t(m)) \in M_2^{n-1}$. Let W_1 be a connected open neighborhood contained inside $\phi_1(M_1^{n-1})$. Define $\sigma(m)$, $m \in \phi_1(M_1^{n-1})$, by $\sigma(m) = -t(\phi_1^{-1}(m))$. Define $\tau(m) = 0$ on the complement of $\bar{W}_1$. Using partitions of unity, we can construct a C^r function $s(m)$ satisfying $s(m) = -t(\phi_1^{-1}(m))$ for $m \in W_1$ and $s(m) = 0$ for m in the complement of $\phi_1^{-1}(M_1^{n-1})$. Let $U_1 = \phi_1^{-1}(W_1)$. Consider $\psi(m) = \alpha(m, s(m))$, $m \in M_2^{n-1}$. $\psi(M_2^{n-1})$ is a differentiable manifold that coincides with M_1^{n-1} on U_1 such that P_2 is topologically equivalent to $\psi(M_2^{n-1}) \times \mathbb{R}$. Relabel $\psi(M_2^{n-1})$ by M_2^{n-1}. We now have $U_1 \subset M_2^{n-1}$.

Repeat the construction for successive P_s^j. We get a sequence $U_1 \subset U_2 \subset \cdots$. Let $M_\infty^{n-1} = \bigcup_i U_i$. It is not difficult to verify that P_∞ is the smallest invariant region containing M_∞^{n-1} and that M_∞^{n-1} is an embedded submanifold of M^n transverse to α. Hence P_∞ is maximal. $\square$

The construction of this lemma is vital to the proofs of many of the following theorems.

2. The Structure of Component of $M^n \backslash S$

2.1. Lemma. Each component C of $M^n \backslash S$ is a parallel region.

2.2. Proposition. If a component C of $M^n \backslash S$ contains a closed orbit γ, then every orbit in C is a closed orbit isotopic to γ.

Proof. Let A be the subset of points of C that lie on closed orbits. Suppose $A \neq C$. Let β be an orbit through a boundary point of A relative to C. Let M_1^{n-1} be an embedded submanifold of C transverse to the flow and $M^{n-1} = M^{n-1} \backslash A$. Let P be the parallel region based on M^{n-1}. The construction of Lemma 1.4 will yield the contradictory fact that β is a separatrix. □

A rather lengthy argument also based on the proof of Lemma 1.4 will give the following proposition.

2.3. Proposition. Let C be a component of $M^n \backslash S$ containing no closed orbits. Then either (i) every submanifold M^{n-1} of C on which C is based is such that every trajectory through points of M^{n-1} intersects M^{n-1} an infinite number of times for both $t > 0$ or $t < 0$, or (ii) there is a submanifold M_0^{n-1} of C, on which C is based, such that every trajectory in C intersects exactly once.

Components C satisfying (i) above will be called types R, (ii) above type H, and Proposition 2.2 type Per. We see that this classifies the components of $M^n \backslash S$. Note that a manifold M^{n-1} generating a region of type H decomposes $C \backslash M^{n-1}$ into two parts C^+ and C^- since C is then topologically equivalent to $M^{n-1} \times \mathbb{R}$.

3. Separatrices and Singular Points

In this section, we will consider isolated singular points of a flow α. If B^n is an isolating disk neighborhood of p, the flow restricted to B^n will be reparametrized so as to be a flow on B^n.

3.1. Definition. A component C of $B^n \backslash S$ will be said to adhere to a singular point p if the singular point is contained in $\overline{C}$.

3.2. Definition. Let C be a region of type H. If there are submanifolds M_1^{n-1} and M_2^{n-1} of C such that C is based on M_1^{n-1} or M_2^{n-1} and $p \in \overline{C_1}^+$, $p \notin \overline{C_1}^-$, $p \in \overline{C_2}^-$, and $p \notin \overline{C_2}^+$, where $C_i^{\pm}$ are the subsets of C referred to at the end of Section 2, then C will be called type HH.

3.3. Definition. Let C be a region of type H. If for any submanifold M^{n-1} on which C is based, $p \in \overline{C}^+$ and $p \in \overline{C}^-$, then C will be called type HE.

3.4. Definition. Let C be a region of type H. If for any submanifold M^{n-1} not adhering to p on which C is based $p \in \overline{C}^+$ and $p \notin \overline{C}^-$, then C will be called type HP_-. Interchanging the roles of $+$ and $-$ gives the definition of type HP_+.

Summarizing the results of the paper thus far, we have:

3.5. Theorem. If p is an isolated singular point of a flow α, a spherical neighborhood B^n of p is decomposed by the separatrix set of the reparametrized flow into the separatrix set $\mathbf{S}$ and regions of type H, Per, R, HH, HE, HP_-, or HP_+ (where H here refers to components of $B^n|\mathbf{S}$ that do not adhere to the origin).

4. Separatrix Manifolds

We consider the following topology on $\mathbf{S}$:

4.1. Definition. Let $\mathcal{N}$ be the collection of open sets in M^n containing no singular points, $\mathcal{B} = \{$connected components of $\mathbf{S} \cap N$ in $M^n : N \in \mathcal{N}\}$, and $\mathcal{T}$ the topology on $\mathbf{S}$ generated by $\mathcal{B}$ and the set $\{\{p\}: p$ is a singular point$\}$.

4.2. Definition. Consider invariant subsets of $K \subset \mathbf{S}$ satisfying

(i) K is pathwise connected in $\mathbf{S}$ relative to $\mathcal{T}$;

(ii) Each point of K is contained in a $\mathcal{T}|K$ open subset of K homeomorphic to an open set of $\mathbb{R}^k$ for some fixed k;

(iii) K contains no singular point if $k > 0$;

(iv) No point x of K is contained in the closure relative to $\mathcal{T}$ of other sets K^1 satisfying conditions (i)–(iii) for the same k;

(v) K is not properly contained in sets satisfying (i)–(iv).

Subsets K of S satisfying (i)–(v) will be called separatrix manifolds.

4.3. Theorem. A separatrix manifold K is a topological manifold on which a differentiable structure can be defined under which the flow α is of class C^r.

REFERENCES

[1] Bhatia, N. P., and L. M. Franklin, Dynamical systems without separatrices, *Funkcial. Ekvac.* **15** (1972), 1–12.
[2] Hartman, P., "Ordinary Differential Equations." Wiley, New York, 1964.
[3] Markus, L., Parallel dynamical systems, *Topology* **8** (1969), 47–57.

Global Results by Local Averaging for Nearly Hamiltonian Systems

JAMES MURDOCK
Department of Mathematics
City College of New York, New York

Consider a space in which the points represent differential equations; a subset consists of Hamiltonian systems, and a further subset, of integrable Hamiltonian systems. Integrable systems do not form an open set, either among Hamiltonian systems or in general, and it is natural to ask what sort of systems lie near an integrable system. For the Hamiltonian case this leads into the theories of Poincarè–Birkhoff and Kolmogorov–Arnol'd–Moser [1, Chapter 4]. We will be concerned with the case of non-Hamiltonian perturbations of an integrable Hamiltonian system. Details are given in [2–4].

According to a general theorem of Arnol'd [1], the phase space of an integrable system is divided up by separatrices into regions admitting action/angle variables r, θ in which the system takes the form $\dot{r} = 0$, $\dot{\theta} = \Omega(r)$. Each torus $r = \text{const}$ is invariant and foliated with minimal invariant tori of lower dimension depending on the commensurability relations satisfied by $\Omega(r)$. Our systems take the form

$$\dot{r} = \varepsilon f(r, \theta)$$
$$\dot{\theta} = \Omega(r) + \varepsilon g(r, \theta) \tag{1}$$

Here $r \in R^n$, $\theta \in R^m$, and f, g are periodic in the components of θ.

According to the method of averaging, since r varies slowly for small ε an approximation to (1) may be constructed by holding r fixed and averaging f over an orbit of the unperturbed system. Defining

$$\bar{f}(r, \theta) = \lim_{T \to \infty} \frac{1}{T} \int_0^T f(r, \theta + \Omega(r)t)\, dt$$

it is easy to show that $\bar{f}$ is the average of f over the minimal invariant torus of the unperturbed system through the point (r, θ). This is a discontinuous function of r because the dimension of the minimal torus through (r, θ) depends on r. A necessary condition for (1) with small ε to have an invariant torus lying near a minimal torus of the unperturbed

25

system is for $\bar{f}$ to vanish on that minimal torus. A global study of (1) therefore requires a study of the zeros of $\bar{f}$.

If $f(r, \theta) = \sum a_\nu(r)e^{i(\nu, \theta)}$, ν a multi-index, then $\bar{f}(r, \theta)$ consists of the terms of the series for which $(\nu, \Omega(r)) = 0$. The set of such ν forms a subgroup of the additive group of multi-indices. The clue to the study of zeros of f is to consider "coarser" averages that are continuous. Namely, if L is a subgroup of the multi-indices, let $\hat{f}_L$ be defined by summing the series for f only over $\nu \in L$. Since L is fixed, $\hat{f}_L$ is a continuous function, and for the "resonance manifold" $\tilde{L}$ of r such that $(\nu, \Omega(r)) = 0$ for all $\nu \in L$, $\hat{f}_L$ is "coarser" than $\bar{f}$, in the sense that $\hat{f}_L$ is obtained by averaging over tori of larger dimension than the minimal tori.

Theorem 1. Zeros of $\bar{f}$ for $r \in \tilde{L}$ can cluster only near zeros of $\hat{f}_L$; in a compact subset of $\tilde{L}$ bounded away from the zeros of $\hat{f}_L$, $\bar{f}$ can only vanish on a finite number of resonance submanifolds of $\tilde{L}$ of lower dimension defined by adjoining to a basis for L additional multi-indices involving small integers.

The idea of the proof is that away from zeros of $\hat{f}_L$ only Fourier coefficients of low order are significant. Theorem 1 provides an inductive procedure for locating the zeros of $\bar{f}$ within small open sets, beginning with the full space of variables (r, θ) and proceeding to submanifolds of lower and lower dimension in a finite number of steps.

The first application is to problems of spin/orbit coupling in the solar system. A spinning planet subject to periodic torque due to its orbit is governed by $\ddot{\theta} = \varepsilon f(t, \theta, \dot{\theta})$, or $\dot{r} = \varepsilon f(t, \theta, r)$, $\dot{\theta} = r$. According to astronomers Mercury and Venus were "despun" by tidal friction until they became "captured" at their present spin rate. Such a capture is impossible unless

$$\bar{f}(t, \theta, r) = \lim_{T \to \infty} \frac{1}{T} \int_0^T f(t, \theta + rt, r)\, dt = 0$$

A coarser average is

$$\hat{f}(r) = \frac{1}{4\pi^2} \int_0^{2\pi} \int_0^{2\pi} f(t, \theta, r)\, dt\, d\theta$$

In the actual example $\hat{f}(r)$ has a single zero, and zeros of $\bar{f}$ can cluster near it, while away from that zero, $\bar{f}$ can vanish only for a few rational values of r. The analysis confirms assumptions made by previous workers, in particular the legitimacy of truncating the Fourier series for f. Small divisor questions can only arise near the zero of $\hat{f}$. For details see [2]. Similar results with quasiperiodic forcing are in [4, Section 6].

Consider a sequence of pendula each coupled to the next by a weak spring, slightly damped, and with the first pendulum in the sequence subject to weak forcing.

Theorem 2. Except for solutions having amplitudes near zero, the steady-state solutions are periodic with period a small integer multiple of the forcing period and with the frequencies of adjacent pendula having small integer ratios. ("Frequency" here is, strictly speaking, frequency of the generating unperturbed solution.)

The integers here are small for the same reason as in Theorem 1. Because of the number of degrees of freedom the inductive procedure is impractical and the proof uses a combination of averaging and energy methods. If the frequency ratio for two adjacent pendula is not a small integer ratio, the energy of the pendula on the side away from the forcing decreases until a resonance capture is formed. The author hopes to apply these ideas to systems of oscillators with higher connectivity, such as entrainment of biorhythms in tissues; however, the "sparseness" of coupling is crucial to the simple result of Theorem 2. Details are given in [4].

REFERENCES

[1] V. I. Arnol'd and A. Avez, "Ergodic Problems of Classical Mechanics." Benjamin, New York, 1968.

[2] J. A. Murdock, Resonance capture in certain nearly Hamiltonian systems, *J. Differential Equations* **17** (1975), 361–374.

[3] J. A. Murdock, Nearly Hamiltonian systems in two degrees of freedom, *Internat. J. Non-Linear Mech.* **10** (1975).

[4] J. A. Murdock, Nearly Hamiltonian systems in nonlinear mechanics: Averaging and energy methods, *Indiana Univ. Math. J.*

Equivalence of Suspensions
and Manifolds with Cross Section*

DAVID R. NAUGLER
Department of Mathematics
Dalhousie University, Halifax, Novia Scotia, Canada

1. Introduction

The notions of cross section of a vector field and suspension of a diffeomorphism, defined in [2], are a bridge between the study of vector fields and that of diffeomorphisms [3].

Let M be a compact smooth manifold with codimension one, submanifold S. If there is a smooth vector field χ on M with S as a cross section, (M, S) is called a cross-section pair, $(M, S; \chi)$ a cross-section triple, χ a cross-sectional vector field, and $X(M, S)$ denotes the C^0 open subset of smooth vector fields that have S as a cross section.

Each $\chi \in X(M, S)$ induces a diffeomorphism of S, denoted $\alpha(\chi)$, by $\alpha(\chi)(x) = \phi_{\tau(x)}(x)$, where $\phi_\tau(x)$ is the flow of χ and $\tau(x)$ the least positive time such that $\phi_{\tau(x)}(x) \in S$. Thus the map $\alpha: X(M, S) \to \text{Diff } S$, where $\text{Diff } S$ is the smooth diffeomorphisms of S, represents each $\chi \in X(M, S)$ by a diffeomorphism, and if $h: M \to \mathbb{R}$ is smooth and positive, $\alpha(\chi) = \alpha(h \cdot \chi)$. Any $\chi \in X(M, S)$ may be smoothly reparameterized (nonuniquely) so that $\tau(x)$, as above, is identically one for the new flow. Such a reparameterized vector field is called normal and induces a smooth bundle structure on M over the circle S^1 with fiber S by the map p defined by $p(\phi_\tau(S)) = \tau + \mathbb{Z}$. Up to bundle equivalence this structure depends only on $\alpha(\chi)$.

Given S and $f \in \text{Diff } S$, a smooth manifold $\text{Sus}(S, f)$, the suspension of f over S, is defined to be the quotient space of $\mathbb{R} \times S$ under the equivalence relation defined by identifying (τ, x) with $(\tau + 1, f^{-1}(x))$. The constant vector field $(1, 0)$ on $\mathbb{R} \times S$ induces a vector field χ on $\text{Sus}(S, f)$ such that χ has cross section $S_0 = \pi(0 \times S)$, where π is the quotient map and S_0 may be identified with S, is normal, and $\alpha(\chi) = f$. The induced bundle structure may be defined intrinsically by $p(\pi(\tau, x)) = \tau + \mathbb{Z}$. Moreover, given $\chi \in X(M, S)$, the bundle structure induced by normalizing χ is bundle equivalent to the bundle structure on $\text{Sus}(S, \alpha(\chi))$.

** These results are contained in the author's thesis, to be submitted to Dalhousie University, under the direction of Professor C. S. Hartzman.

2. Equivalence of Suspensions

The natural types of equivalence for cross-section pairs and triples are equivalence as bundles (B-equivalence) and equivalence as pairs (equivalence). By the above remarks, it is enough to consider suspensions.

Two diffeomorphisms f, g on S are said to be conjugate (pseudo-) isotopic iff f is (pseudo-) isotopic to hgh^{-1} for some $h \in$ Diff S.

Theorem 2.1. Sus(S, f) is B-equivalent to Sus(S, g) if and only if f is conjugate isotopic to g.

It follows that the map $f \to$ Sus(S, f) induces a one-to-one correspondence between similarity classes of the group of isotopy components of Diff S, and B-equivalence classes of suspensions over S.

Theorem 2.2. Sus(S, f) is equivalent to Sus(S, g) if and only if f is conjugate pseudoisotopic to g or g^{-1}.

As above, a classification result follows.

As an example, consider the torus $T^2 = \mathbb{R}^2/\mathbb{Z}^2$. The group of (pseudo-) isotopy components of Diff T^2 is isomorphic to $GL(2, \mathbb{Z})$, the 2×2 integer matrices with determinant ± 1, by the map induced by $A \to \tilde{A}$, where $\tilde{A}$ is the diffeomorphism of T^2 induced by $A \in GL(2, \mathbb{Z})$ considered as a linear map of $\mathbb{R}^2$. The inverse of this isomorphism may be considered to be induced by the map taking f into the induced isomorphism on $\pi_1 T^2 \approx \mathbb{Z}^2$. Given the trace and determinant, there are a finite number of similarity classes in $GL(2, \mathbb{Z})$ with these invariants (unless the trace is ± 2 and the determinant is one), representatives of which may be enumerated (with some redundancy). Other similarity class invariants may be used to limit the possibilities further, so that the classification of suspensions over the torus is almost complete.

3. Manifolds with Cross Section

Given cross-section triples $(M, S; \chi)$ and $(N, T; \psi)$ that have induced maps f and g, respectively, it is known [2] that f topologically (smoothly) conjugate to g implies that χ is topologically (smoothly) equivalent to ψ. The converse, which is not always true, is known to hold [1] when $\pi_1 S$

has no element of infinite order. The following more general result puts the restriction on M.

Note that the first homology group of a manifold with a cross section has rank at least one.

Theorem 3.1. If $H_1(M, \mathbb{Z})$ has rank precisely one, then χ is topologically (smoothly) equivalent to ψ if and only if f is topologically (smoothly) conjugate to g.

Corollary 3.2. Let $H_1(M, \mathbb{Z})$ have rank precisely one. $\chi, \psi \in X(M, S)$ are topologically (smoothly) equivalent if and only if $\alpha(\chi)$ is topologically (smoothly) conjugate to $\alpha(\psi)$.

As examples consider suspensions over the torus. It may be shown that $H_1(\mathrm{Sus}(T^2, \tilde{A}), \mathbb{Z})$ has rank one unless the trace of A is 2(0) and the determinant of A is one (minus one). Indeed, for any $n \geq 2$ and any finite abelian group G with at most n generators, there is an $A \in GL(n, \mathbb{Z})$ such that $H_1(\mathrm{Sus}(T^n, \tilde{A}), \mathbb{Z})$ is isomorphic to $\mathbb{Z} \oplus G$.

REFERENCES

[1] G. Ikegami, On classification of dynamical systems with cross sections, *Osaka J. Math.* **6** (1969), 419–433.
[2] S. Smale, Stable manifolds for differential equations and diffeomorphisms, *Ann. Scuola Norm. Sup. Pisa* **17**(3) (1963), 97–116.
[3] S. Smale, Differentiable dynamical systems, *Bull. Amer. Math. Soc.* **73** (1967), 747–817.

Structural Stability Theorems*

R. CLARK ROBINSON
Department of Mathematics
Northwestern University, Evanston, Illinois

A C^1 (time independent) flow is a function $f: R \times M \to M$ such that it has continuous first partial derivatives, for each $t \in R$ $f^t: M \to M$ is a diffeomorphism, and it satisfies the group property, $f^t \circ f^s(x) = f^{t+s}(x)$. The manifold M is assumed to be compact and without boundary. A neighborhood N of f in the set of C^1 flows is those g such that g restricted to $[0, 1] \times M$ is uniformly near f and all the first partial derivatives are uniformly near those of f. We want to find those flows whose orbit structure is stable under perturbations. We can ask the same questions about diffeomorphisms. This study of diffeomorphisms is the same as that for periodic time-dependent flows. Again a neighborhood N of f is those g that are uniformly near f pointwise and whose first partial derivatives are uniformly near those of g.

We say g is *conjugate* to f if there exists a homeomorphism taking oriented trajectories of f to oriented trajectories of $g : h \circ f^{\alpha(t, x)}(x) = g^t \circ h(x)$, where α is a reparameterization of f and $h: M \to M$ is a homeomorphism. If f and g are diffeomorphisms we have the same definition but do not allow any reparameterization. We say a flow f (or diffeomorphism) is *structurally stable* if there is a neighborhood N of f in C^1 flows (or C^1 diffeomorphisms) such that every $g \in N$ is conjugate to f. Thus f is structurally stable if every g near f has the same orbit structure as f.

We want to define conditions that imply structural stability. If we look at a periodic point of a diffeomorphism, to get stability we can not have a center, i.e., the eigenvalues of $Df^p(x)$ can not be of absolute value one, where $f^p(x) = x$. This is a hyperbolicity assumption on the periodic points. In two dimensions a stable diffeomorphism can have a sink, source, or saddle but not a rotation or saddle node.

---

* Research for this chapter was done at Northwestern University, and was partially supported by NSF Grant GP 19815. [AMS classification 58F10.]

We need hyperbolicity on more than just the periodic points. The *nonwandering set* Ω of f is the set of points $x \in M$ such that for all neighborhoods U of x there is a $T \geq 1$ such that $U \cap f^t(U) \neq \varnothing$. The nonwandering set contains all the α and ω limit points and has a general type of recurrence property. We say f *satisfies Axiom A* if (Aa) f satisfies a hyperbolicity assumption on Ω, and (Ab) the periodic points are dense in Ω. More precisely, (Aa) the tangent space to M over Ω splits into two continuous subbundles E^s and E^u, called the stable and unstable bundles, and there are constants $C > 0$ and $0 < \lambda < 1$, such that for $t \geq 0$

$$|Df^t(x)V_x^s| \leq C\lambda^t |V_x^s| \qquad \text{for} \quad V_x^s \in E_x^s$$
$$|Df^{-t}(x)V_x^u| \leq C\lambda^t |V_x^u| \qquad \text{for} \quad V_x^u \in E_x^u$$

For a flow there are three subbundles, E^s, E^u, and a bundle spanned by $(d/dt)f^t(x)$.

An example of a hyperbolic set occurs when there is a homoclinic point for a saddle fixed point. Let p be a saddle point of a diffeomorphism in the plane, let

$$W^s(p) = \{x \in M: d(f^t(x), f^t(p)) \to 0 \text{ as } t \to \infty\}$$

be the stable manifold, and let

$$W^u(p) = \{x \in M: d(f^t(x), f^t(p)) \to 0 \text{ as } t \to -\infty\}$$

be the unstable manifold. They are both immersed curves in the plane. A *homoclinic point* is a transversal intersection of these two curves $y \in W^s(p) \cap W^u(p)$ with $y \neq p$. If we take a long box B along the stable manifold, a high iterate of B by f crosses B twice, $f^T(B) \cap B = B_1 \cup B_2$ (Fig. 1). The intersection $\bigcap_{n=-\infty}^{\infty} f^{nT}(B)$ contains a cantor set with periodic

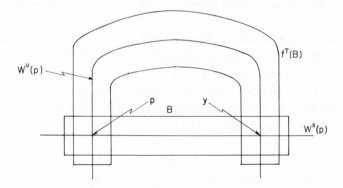

FIG. 1.

points dense. The box is stretched in the vertical direction and contracts in the horizontal direction, so these are the unstable and stable bundles (see [12]).

If f satisfies Axiom A, then the stable and unstable manifolds of all orbits are immersed submanifolds. If two of these intersect, there can not be a tangency because this is an unstable phenomenon. We say f satisfies the *strong transversality condition* if all stable manifolds of orbits that meet another unstable manifold of an orbit do so transversally. We can now state the theorem.

Theorem. Let f be a C^1 flow or diffeomorphism that satisfies Axiom A and the strong transversality condition. Then f is structurally stable.

Peixoto proved that for flows on compact two-manifolds, these are necessary and sufficient conditions for structural stability [7]. Anosov and Moser proved the theorem when $M = \Omega$ [1, 5]. When Ω is a finite union of orbits, it is proved by Palis and Smale [6]. Robbin proved the case when f is a C^2 diffeomorphism [8]. The author adapted this proof for C^2 vector fields and more recently has proved it for C^1 diffeomorphisms and flows [9–11].

Some progress has been made on the converse by Franks and Guckenheimer [2–4], and just recently there is the following result.

Theorem (Mañé). If dim M is two and f a structurally stable diffeomorphism, then it satisfies Axiom A and the strong transversality condition.

Now we give an idea of the proof of structural stability contained in [10] or [11]. We use the stable manifold approach of Anosov, the induction process of Palis, and the d_f metric introduced by Robbin. First we look at stability near a saddle fixed point p for a diffeomorphism in the plane. This is related to the linearization result of Hartman. For g near f, we construct a family of disks $\{Z^u(x) : x \in U\}$, where U is a neighborhood of p such that (1) $Z^u(x)$ is a disk near x in approximately the unstable direction, (2) g restricted to $Z^u(x)$ is an expansion, and (3) g takes the disk near x to the disk near $f(x)$, $g(Z^u(x)) \supset Z^u(f(x))$. Then $\{Z^u(x) : x \in U\}$ is a disk bundle over U. Since g^{-1} is a fiber contraction, we can construct a fixed section $h(x) \in Z^u(x)$ such that $g^{-1}h(x) = hf^{-1}(x)$ or $gh(x) = hf(x)$. To prove that h is one to one, we show h is d_f Lipschitz near the identity. The hardest part of the analysis goes into proving this. (For Ω finite this can be avoided.)

When Ω is more complicated it breaks up into a finite number of pieces that are indecomposable, $\Omega = \Omega_1 \cup \cdots \cup \Omega_m$. Then perhaps reordering the Ω_i, it is possible to get a Lyapunov function $L: M \to R$ such that $\Omega_i \subset L^{-1}(i)$ and $L \circ f^t(x) > L(x)$ for $x \notin \Omega$ and $t > 0$. [For a flow $(d/dt)L \circ f^t(x) > 0$ for $x \notin \Omega$.] The conjugacy we construct must be defined compatibly near the different Ω_i. Palis was the first to overcome this difficulty. For g near f, we construct compatible families of unstable disks $\{Z_i^u(x) : x \in \mathcal{O}^f(U_i)\}$, where U_i is a neighborhood of Ω_i and $\mathcal{O}^f(U_i)$ is the orbit of U_i by f, such that (1) each $Z_i^u(x)$ is a disk near x of dimension equal to $\dim(E^u|\Omega_i)$; (2) for $x \in U_i$, $Z_i^u(x)$ is approximately in the unstable direction and g restricted to $Z_i^u(x)$ is an expansion; (3) $g(Z_i^u(x)) \supset Z_i^u(f(x))$; and (4) (compatibility) if $x \in \mathcal{O}^f(U_i) \cap \mathcal{O}^f(U_j)$ with $i < j$, then $Z_i^u(x) \supset Z_j^u(x)$.

Once these families are constructed, then again g^{-1} is a contraction for x near Ω so we can construct an invariant section. Again we need to show h is d_f Lipschitz near the identity, where $d_f(x, y) = \sup\{d(f^n(x), f^n(y)) : n \in Z\}$.

REFERENCES

[1] D. Anosov, Geodesic flows on closed Riemannian manifolds, *Proc. Steklov Inst.* **90** (1967) (Amer. Math. Soc. transl., 1969).

[2] J. Franks, Differentiably Ω-stable diffeomorphisms, *Topology* **11** (1972), 107–113.

[3] J. Franks, Time dependent stable diffeomorphisms, *Invent. Math.* **24** (1974), 163–172.

[4] J. Guckenheimer, Absolutely Ω-stable diffeomorphisms, *Topology* **11** (1972), 195–197.

[5] J. Moser, On a theorem of Anosov, *J. Differential Equations* **5** (1969), 411–440.

[6] J. Palis and S. Smale, Structural stability theorems, *Proc. Symp. Pure Math. Soc.* **14** (1970), 223–232.

[7] M. Peixoto, Structural stability on two dimensional manifolds, *Topology* **1** (1962), 101–120.

[8] J. Robbin, A structural stability theorem, *Ann. of Math.* **94** (1971), 447–493.

[9] R. C. Robinson, Structural stability of vector fields, *Ann. of Math.* **99** (1974), 154–175.

[10] R. C. Robinson, Structural stability of C^1 flows, *Proc. Conf. Dynam. Syst. Univ. of Warwick 1974, Lecture Notes in Math.* pp. 262–277. Springer-Verlag, Berlin and New York, 1975.

[11] R. C. Robinson, Structural Stability of C^1 Diffeomorphisms, mimeographed paper, Northwestern Univ. (1974).

[12] S. Smale, Differentiable dynamical systems, *Bull. Amer. Math. Soc.* **73** (1967), 747–817.

Numerical Studies of an Area-Preserving Mapping*

MARTIN BRAUN
Lefschetz Center for Dynamical Systems
Division of Applied Mathematics
Brown University, Providence, Rhode Island

Consider the following mapping M of the plane onto itself. Let P_1 and P_2 be two points separated by a small distance ε. Let M_1 denote the mapping in which each point is rotated about P_1 in a counterclockwise direction by an angle equal to the reciprocal of its distance from P_1 raised to the $\frac{4}{3}$ power. Let M_2 be the corresponding rotation about P_2. The mapping M is defined to be the composition of M_1 with M_2. Clearly, M is an area-preserving mapping of the plane onto itself.

This mapping arose from the author's study of the Störmer problem, which is the motion of a charged particle in the earth's magnetic field. The motion of such a charged particle is governed by the Hamiltonian

$$H = \frac{1}{2}\left(p_\rho{}^2 + p_z{}^2\right) + \frac{1}{2}\left(\frac{1}{\rho} - \frac{\rho}{r^3}\right)^2, \qquad r^2 = \rho^2 + z^2, \tag{1}$$

where $z = 0$ is the equatorial plane. It has been shown [1] that the equatorial plane is a surface of section for the flow governed by the Hamiltonian (1). In a typical orbit [2] the particle crosses the equatorial plane in the northerly direction; it penetrates to a certain latitude and then gets reflected back, penetrating to a certain latitude below the equator. If we eliminate p_z on each surface of constant energy

$$E = \frac{1}{2}\left(p_\rho{}^2 + p_z{}^2\right) + \frac{1}{2}\left(\frac{1}{\rho} - \frac{\rho}{r^3}\right)^2$$

and numerically compute the section map Σ taking the $\rho - p_\rho$ plane into itself, we observe the following: If the particle does not penetrate to very high latitudes on either side of the equator, then its orbit cuts the surface of section in a sequence of points all of which seem to lie on closed invariant curves. It has been shown [3] that these curves correspond to the particles trapped in the Van Allen radiation belt. On the other hand, if

* This research was supported in part by the Air Force Office of Scientific Research under AFOSR-71-2078C and in part by the National Science Foundation under GP-28931X2.

the particle penetrates to very high latitudes, i.e., it comes near the singularity $r = 0$, then its orbit cuts the surface of section in a very "fuzzy" sequence of points. We would like to conclude that the invariant curves have disintegrated in this region, but numerical errors preclude any definitive conclusion.

In order to determine the qualitative nature of the section map Σ for those orbits that penetrate close to the singularity, we must first determine the nature of the flow near the singularity $r = 0$. It has been shown [4] that near the singularity, for $z > 0$, the flow on each constant energy surface is topologically equivalent to the orbits of the system of differential equations

$$\mathring{q} = r^{7/3}, \qquad\qquad y = y_1 + iy_2, \qquad h \text{ real.} \qquad (2)$$
$$\mathring{y} = ih(q, y, \bar{y})y, \qquad r^2 = q^2 + |y|^2,$$

Exactly the same result is true for $z < 0$. Notice that each cylinder $y_1^2 + y_2^2 = c$ is invariant under the flow. The image of these cylinders in the ρ, z, p_ρ, p_z phase space will intersect the equatorial place in closed curves when continued forward under the flow. Now the orbit $y = 0$ corresponds to the unique orbit entering the singularity from the north ($q < 0$), and leaving the singularity ($q > 0$). If this orbit crosses the equatorial plane at right angles, then it must enter the singularity from the south, since the Hamiltonian (1) is a function of z^2. Moreover, the cylinders from the north and the cylinders from the south would join up exactly, and the Störmer problem would be integrable. However, the orbit leaving the singularity from the north just misses crossing the equatorial plane at right angles, and this error is measured by ε. The mapping M_1 is a model for the flow when the particle starts in the equatorial plane, goes north, and then comes back, while the mapping M_2 is a model for the flow when the particle starts in the equatorial plane, goes south, and then returns to the equatorial plane. Hence, the mapping $M = M_2 \circ M_1$ is a model for the section map Σ.

It can be rigorously proven [5, 6] that the mapping M possesses infinitely many invariant curves for r sufficiently large. This is a simple consequence of the Moser twist theorem [7]. In addition, there is a region near the points P_1 and P_2 in which the mapping is equivalent to a horseshoe mapping. Consequently, M also possesses infinitely many homoclinic points. Numerical studies of M show that the flow is transitive in this neighborhood. In addition, numerical studies provide a nice explanation of how invariant curves disintegrate and then form anew. To wit, numerical results thus far clearly indicate that the stable and unstable curves, which in the integrable case emanate from the hyperbolic periodic points of a mapping M and surround the islands of closed curves containing the elliptic periodic points, are replaced in the nonintegrable case by very narrow bands in which the mapping is ergodic. These bands connect the numerous hyperbolic periodic points to each other, and provide a very natural description of the disintegration and formation anew of closed invariant curves for the mapping M.

REFERENCES

[1] DeVogelaere, R., L'equation de Hill et le problème de Störmer, *Canad. J. Math.* **2** (1950), 440.

[2] Störmer, C., "The Polar Aurora." Oxford Univ. Press (Clarendon), London and New York, 1955.

[3] Braun, M., Particle motions in a magnetic field, *J. Differential Equations* **8** (1970), 294–332.

[4] Braun, M., Structural stability and the Störmer problem *Ind. Univ. Math. J.* **20** (1970), 469–497.

[5] Braun, M., Some recent results for the Störmer problem (in preparation).

[6] Braun, M., Some theoretical and numerical results of an area preserving mapping (in preparation).

[7] J. Moser, On invariant curves of area-preserving mappings of an annulus, *Nachr. Akad. Wiss., Göttingen: Math. Phys. Kl.* **IIa** (1) (1962), 1–20.

A Geometrical Approach to Classical Mechanics

CHONG-PIN ONG†
Department of Mathematics
University of California, Berkeley, California

1. Introduction

In Riemannian geometry one studies "natural paths" called geodesics of a Riemannian manifold (M, g). The behavior of the geodesics is related to the Riemannian curvatures of the manifold.

In our approach to classical mechanics we exploit the above relationship between geodesics and curvatures. The setting is a simple mechanical system (M, K, V), where M is a smooth manifold that represents the configuration space where the motion takes place. The tangent bundle of M, TM, is the phase-space. The kinetic energy of the system $K: TM \to R$ is derived from a Riemannian metric g on M by

$$K(v) = \tfrac{1}{2}g(v, v) \qquad \text{for all tangent vectors } v.$$

Finally, the potential energy function $V: M \to R$ is simply a smooth function on M. The "natural paths" of a simple mechanical system, called *physical paths*, are those paths $\gamma(t)$ in M that satisfy Newton's law of motion:

$$(D/dt)\gamma'(t) = -\operatorname{grad} V,$$

where D/dt is the covariant derivative and $\operatorname{grad} V$ the gradient of V relative to the given metric g.

The physical paths may be regarded as geodesics as follows:

Theorem (Jacobi). A path γ in M is a physical path of total energy value $h \Leftrightarrow \gamma$ is a geodesic relative to the metric $g_h = 2(h - V)g$.

We call g_h an *h-Jacobi metric*. Note that the total energy $E = K + V$ is conserved along a physical path.

Our task is to investigate the Riemannian curvatures relative to the h-Jacobi metric g_h. We shall call these the *h-mechanical curvatures* of the

---

† Present address: Department of Mathematics, Iowa State University, Ames, Iowa.

simple mechanical system relative to the total energy value h. In this chapter we give a summary of some of the results on the study of these h-mechanical curvatures. A complete exposition is found in [4].

2. Statement of Results

We shall denote h-mechanical curvature quantities by an asterisk. For example, Ric* shall denote the h-mechanical Ricci curvature, while Ric is the Ricci curvature relative to the kinetic energy metric g, etc. For the following results we assume that the boundary ∂M_h of the admissible configuration manifold $M_h = V^{-1}(-\infty, h)$ is nonempty. In fact, ∂M_h is a submanifold of codimension 1 if h is a regular value of V.

Theorem A.   As $x \to \partial M_h$:

(i)   For $n \le 3$, $\text{Ric*}(x)(\xi, \xi) \to \infty$ for all tangent vectors $\xi \in T_x M$, $\xi \ne 0$.

(ii)   For $n = 4$, $\text{Ric*}(\xi, \xi) \to \infty$ for all $\xi \in T_x M$, ξ not orthogonal to grad V, i.e., $dV(\xi) \ne 0$.

(iii)   For $n > 4$, $\text{Ric*}(x)(\xi, \xi) \to \infty$ if $\xi \ne 0$ is within the cone defined by the angle $\alpha = \cos^{-1}(1/\sqrt{3})$ from grad V as the axis of the cone; and $\text{Ric*}(x)(\xi, \xi) \to -\infty$ if ξ is outside the cone defined by $\beta = \cos^{-1}(1/3)$ from grad V (Fig. 1).

Let $K^*(x)(\pi)$ be the h-mechanical sectional curvature, where π is a two-plane in $T_x M$; $\pi = \{\xi, \eta\}$.

Theorem B.   As $x \to \partial M_h$:

(i) $K^*(\pi) \to \infty$ if π contains a vector ξ interior to the cone defined by the angle $\cos^{-1}(1/\sqrt{3})$ from grad V as the axis.

(ii)   Suppose π contains a vector orthogonal to grad V, then $K^*(\pi) \to -\infty$ if the plane π is strictly outside the cone as in (i).

Remark.   In particular, as $x \to \partial M_h$,

$$K^*(\pi) \to \infty \qquad \text{if} \quad \pi \text{ contains grad } V,$$
$$K^*(\pi) \to -\infty \qquad \text{if} \quad \pi \text{ is orthogonal to grad } V.$$

These results are obtained by computing and examining the formulas

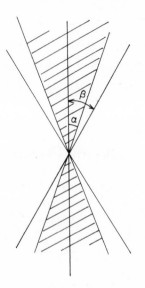

FIG. 1

of the h-mechanical curvatures. Note that the h-Jacobi metric is just a conformal change of the Riemannian metric g.

With the following theorem we may apply the above results to the three-body problem.

Theorem 1. Except for masses $(m_1, m_2, m_3) \in R^+ \times R^+ \times R^+$, in the algebraic surface in $R^+ \times R^+ \times R^+$ defined by $(m_3/m_1)^{1/2} + (m_3/m_1)^{1/2} = 1$ any total energy value h is a regular value of the Newtonian potential function V.

In the next main result, we find m "curvature functions" $Q_1, \ldots, Q_m$, defined globally on M_h. Here $m = \binom{n}{2}$, where $n = \dim M$. These curvature functions determine the sign of the h-mechanical sectional curvature as follows.

Theorem C

(i) If $K(x)\pi \geq 0$ and the curvature functions $Q_i(x) \geq 0$ for $i = 1, \ldots, m$, then $K^*(x)\pi \geq 0$.

(ii) If $K(x)\pi \leq 0$ and $Q_i(x) \leq 0$ for i odd and $Q_j(x) \geq 0$ for j even, then $K^*(x)\pi \leq 0$.

(iii)   If $K(x)\pi = 0$, then

$$K^*(x)\pi \geq 0 \Leftrightarrow Q_i(x) \geq 0 \qquad \text{for all} \quad i = 1, \ldots, m,$$
$$K^*(x)\pi \leq 0 \Leftrightarrow Q_i(x) \leq 0 \qquad \text{for} \quad i \text{ odd,}$$
$$\Leftrightarrow Q_j(x) \geq 0 \qquad \text{for} \quad j \text{ even.}$$

3. Construction of the Curvature Functions

The h-mechanical sectional curvature $K^*(\pi)$ is given by

$$K^*(\pi) = \frac{1}{2(h - V)} \left[K(\pi) - \frac{\langle Q_{XY} X, Y \rangle}{\| X \wedge Y \|^2} \right],$$

where $\pi = \{X, Y\}$, and

$$\langle Q_{XY} X, Y \rangle = \frac{1}{n - 2} [2C(X, Y)\langle X, Y \rangle - C(X, X)\| Y \|^2 - C(Y, Y)\| X \|^2].$$

Note that $g(X, Y)$ is often written $\langle X, Y \rangle$. Finally, C is the fundamental symmetric two-tensor defined in terms of the potential function V by

$$C(X, Y) = \frac{n - 2}{4(h - V)^2} [2(h - V) \text{ Hess } V(X, Y)$$
$$+ 3 \, dV(X) \, dV(Y) - \| \text{grad } V \|^2 \langle X, Y \rangle],$$

where Hess $V(X, Y) = \langle \nabla_X \text{ grad } V, Y \rangle$.

For each $x \in M_h$, the symmetric two-tensor $C(x)$ can be identified with a self-adjoint linear operator $\tilde{C}(x): T_x M_h \rightarrow T_x M_h$ via the metric $\langle \cdot, \cdot \rangle$ as follows:

$$C(x)(v, w) = \langle \tilde{C}(x)v, w \rangle \qquad \text{for all} \quad v, w \in T_x M_h.$$

Let $\tilde{C}_U$ and $\tilde{C}_V$ be the local representation of $\tilde{C}$ in the coordinate neighborhoods U and V, and ρ the coordinate transformation. Then because $C(x)$ is a two-tensor the following diagram commutes:

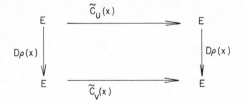

i.e., $D\rho(x)\tilde{C}_U(x) D\rho(x)^{-1} = \tilde{C}_V(x)$. So if φ is an invariant polynomial in $\mathcal{Gl}(n)$ = Lie algebra of the $n \times n$ matrices, i.e., $\varphi(X) = \varphi(AXA^{-1})$ for all $A \in Gl(n)$ and $X \in \mathcal{Gl}(n)$, then $\varphi(\tilde{C}(x))$ is well defined.

For $X \in \mathcal{Gl}(n)$, let $p_i(X)$, $i = 1, \ldots, n$, be the elementary invariant polynomials, i.e.,

$$\det(z - X) = z^n - p_1(X)z^{n-1} + \cdots + (-1)^n P_n(X).$$

We define the *elementary curvature functions*

$$P_i(x) = p_i(\tilde{C}(x)) \qquad \text{for} \quad i = 1, \ldots, n.$$

It will be apparent below that the following proposition is true.

Proposition.   The curvature functions $Q_1(x), \ldots, Q_m(x)$ belong to the ring generated by the elementary curvature functions, i.e.,

$$Q_1(x), \ldots, Q_m(x) \in R[P_1(x), \ldots, P_n(x)].$$

To define these curvature functions, consider a self-adjoint linear operator $A: E \to E$. Let $\lambda_1, \ldots, \lambda_n$ be its eigenvalues. Let $p_1(\lambda_1, \ldots, \lambda_n)$ be the elementary symmetric polynomials, i.e.,

$$p_i(\lambda_1, \ldots, \lambda_n) = \sum_{s_1 < s_2 < \cdots < s_i} \lambda_{s_1} \lambda_{s_2} \cdots \lambda_{s_i}.$$

Observe that $p_i(\lambda) = p_i(A)$, the elementary invariant polynomial. Next consider the polynomial

$$Q(z) = \prod_{i < j}(z - (\lambda_i + \lambda_j))$$

$$= z^m - q_1(\lambda_1, \ldots, \lambda_n)z^{m-1} + \cdots + (-1)^m q_m(\lambda_1, \ldots, \lambda_n)$$

where $m = \binom{n}{2}$. Note that $q_i(\lambda)$ are all symmetric polynomials and so

$$q_i(\lambda) \in R[p_1(\lambda), \ldots, p_n(\lambda)].$$

Since $p_i(\lambda) = p_i(A)$, we may write $q_i(\lambda) = q_i(A)$ and conclude that $q_i(A)$ are invariant polynomials. Thus $q_i(\tilde{C}(x))$ are globally defined. Finally, the *curvature functions* are defined by

$$Q_i(x) = q_i(\tilde{C}(x)), \qquad i = 1, \ldots, m.$$

REFERENCES

[1] Kobayashi, S., and Nomizu, K., "Foundations of Differential Geometry," Vols. I and II. Wiley (Interscience), New York, 1969.

[2] Laugwitz, D., "Differentialgeometrie." Teubner, Stuttgart, 1960.
[3] Milnor, J., "Topology from the Differentiable Viewpoint." Univ. Press of Virginia, Charlottesville, Virginia, 1965.
[4] Ong, C. P., Curvature and mechanics, *Advan: Math.* **15** (3) (1975).
[5] Ong, C. P., Curvature of the Lagrange Relative Equilibrium. Preprint.
[6] Pollard H., "Math Introduction to Celestial Mechanics." Prentice Hall, Englewood Cliffs, New Jersey, 1966.
[7] Smale, S., Topology and mechanics I, *Invent. Math* **10** (1970).
[8] Smale, S., Topology and Mechanics II, *Invent. Math.* **11** (1971).
[9] Yano, K., and Bochner, S., "Curvature and Betti Numbers" (Ann. Math. Stud. Vol. 32). Princeton Univ. Press, Princeton, New Jersey, 1953.

Chapter 2 : GENERAL THEORY

A Solution of Ulam's Conjecture on the Existence of Invariant Measures and its Applications

ANDRZEJ LASOTA
Department of Mathematics
Jagellonian University, Krakow, Poland

1. Rényi Theorem

Assume that the system under consideration is governed by the equation

$$x_{n+1} = \tau_r(x_n),$$ (1)

where τ is a transformation of the interval $[0, 1]$ into itself, given by the formula

$$\tau_r(x) = rx \pmod 1, \qquad r > 1.$$ (2)

We interpret the index n as time. Given a function $\varphi \in L_1[0, 1]$ and an initial point x_0, it is possible to formulate the problem of seeking the mean value of the function φ along the trajectory

$$\overline{\varphi}(x_0) = \lim_n \frac{1}{n} \sum_{k=1}^{n} \varphi(x_k).$$ (3)

From the individual Birkhoff theorem, it follows that for almost all x_0

$$\overline{\varphi}(x_0) = \int_0^1 \varphi(x)\mu(dx),$$ (4)

where μ is an invariant ergodic measure on $[0, 1]$.* So the problem of finding the mean value (3) becomes a problem of finding an invariant measure. For transformation (2), this problem was not solved until the year 1957. It was then that Rényi [8] showed that for every $r > 1$ there exists exactly one invariant measure with respect to τ_r equivalent to Lebesgue measure. In the sequel we will show how, in some technical problems,

---

* We assume that all measures we consider are nonnegative and normalized.

there appear transformations somewhat more general than (2), and how it is possible to find invariant measures for them.

2. Kryloff–Bogoliuboff Theorem

For transformations $\tau\colon X \to X$ continuous on a topological space X, the problem of the existence of an invariant measure was solved in 1937 by Kryloff and Bogoliuboff. According to their theorem, such a measure defined on a σ-algebra of Borel sets exists when the topological space X is compact. Let us see what this means in the case we are discussing. Consider transformation (2) for $r = 2$. This transformation is continuous on the torus $\{x \in \mathbb{R}(\mathrm{mod}\ 1)\}$ and according to the Kryloff–Bogoliuboff theorem there exists an invariant measure. Such a measure may be, for instance,

$$\mu(A) = \begin{cases} 1, & 0 \in A, \\ 0, & 0 \notin A. \end{cases} \tag{5}$$

Nevertheless, the ergodic theorem applied to this measure says nothing about the behavior of value $\overline{\varphi}(x_0)$ on the set of μ-measure zero, i.e., for all $x_0 \in (0, 1)$. A set of full measure consists of only one point $x_0 = 0$, and for the trajectory starting from this point there is, of course, $x_0 = x_1 = x_2 = \cdots$ and, consequently,

$$\overline{\varphi}(x_0) = \frac{1}{n} \sum_{k=0}^{n-1} \varphi(x_k) = \varphi(x_0).$$

This is a rather banal result. Generally, we can say that the Kryloff–Bogoliuboff theorem becomes banal if transformation τ has a fixed point. The measure can be concentrated only on this point. So there arises a natural problem of seeking some more interesting measures. For the interval $[0, 1]$ it is advantageous to find a measure absolutely continuous with respect to the Lebesgue measure. The support of such a measure has, of course, a positive Lebesgue measure, and consequently the ergodic theorem describes the behavior of $\overline{\varphi}(x_0)$ on a more interesting set. Besides, an absolutely continuous measure also has the additional advantage that every bounded measurable function is integrable with respect to this measure, so it is possible to apply the ergodic theorem to a rather broad class of functions.

3. Ulam's Conjecture

It is easy to see that for the transformation $\tau: [0, 1] \to [0, 1]$ there does not exist an absolutely continuous invariant measure if the graph of τ is too flat ($|\tau'|$ is too small). For transformation (2) such measure does not exist if $|r| < 1$.

Ulam [9] posed the problem of the existence of an absolutely continuous invariant measure for the transformation defined by a sufficiently simple function (e.g., a broken line function or a polynomial) the graph of which does not cross the line $y = x$ with a slope of absolute value less than 1. The literal answer for this question is, in fact, negative, namely, we can show that an absolutely continuous measure invariant under transformation τ given by the formula

$$\tau(x) = \begin{cases} 1 - 2x, & 0 \le x \le 5/12 \\ (2/7)(1 - x), & 5/12 \le x \le 1, \end{cases} \tag{6}$$

does not exist. Transformation (6) gives us a continuous broken line and crosses the line $y = x$ at the point $x = \frac{1}{3}$ with slope $\tau'(1/3) = -2$.

However, if we simply admit a slightly stronger assumption that

$$\inf \left| \frac{d}{dx} \tau^N(x) \right| > 1, \tag{7}$$

the answer for Ulam's problem for the function piecewise of class C^2 is positive. Let us remember that a transformation $\tau: [0, 1] \to R$ is called *piecewise* C^2, if there exists a partition $0 = a_0 < a_1 < \cdots < a_p = 1$ of the unit interval such that for each integer i $(i = 1, \ldots, p)$ the restriction τ_i of τ to the open interval (a_{i-1}, a_i) is a C^2 function that can be extended to the closed interval $[a_{i-1}, a_i]$ as a C^2 function. τ need not be continuous at the points a_i. Then we have the following [5, 11]

Theorem 1. Let $\tau: [0, 1] \to [0, 1]$ be a piecewise C^2 function satisfying condition (7) for a positive integer N. Then there exists an absolutely continuous measure invariant under τ.

Measure μ in Theorem 1 does not need to be unique. A simple criterion of uniqueness follows from the interesting result of Li and Yorke [6], who proved that under the assumptions of Theorem 1 there exists a sequence of absolutely continuous invariant measures

$$\mu_1, \ldots, \mu_q, \qquad q \le p - 1,$$

such that every absolutely continuous measure invariant under τ can be shown as a convex combination

$$\mu = \sum_{i=1}^{q} \lambda_i \mu_i, \quad \lambda_i \geq 0, \quad \sum_{i=1}^{q} \lambda_i = 1. \tag{8}$$

So, in particular, for $p = 2$, which means that the transformation τ consists of two pieces of class C^2, the measure μ is unique.

4. Frobenius–Perron Operator

Now we shall discuss the problem of constructing the invariant measure mentioned in Theorem 1. For this, it is convenient to consider the deterministic process

$$x_{n+1} = \tau(x_n) \tag{9}$$

as a particular case of a Markoff process. We will understand as a Markoff process (in Hopf's sense) a linear operator $P: L_1[0, 1] \rightarrow L_1[0, 1]$ that satisfies the following two conditions:

(a) $f \geq 0 \Rightarrow Pf \geq 0$
(b) $f \geq 0 \Rightarrow \|Pf\|_{L_1} = \|f\|_{L_1}$.

Given the transformation $\tau: [0, 1] \rightarrow [0, 1]$ we define the respective Markoff process by the formula

$$P_\tau f(x) = \frac{d}{dx} \int_{\tau^{-1}[0, x]} f(s)\, ds. \tag{10}$$

It is easy to check that the operator P_τ satisfies conditions (a) and (b). From definition (10) it also follows that the absolutely continuous measure μ is invariant if its density $d\mu/dx = f$ satisfies the equation

$$f = P_\tau f. \tag{11}$$

The problem of constructing the invariant measure is, in fact, the problem of seeking approximate solutions of Eq. (11). Such a solution can be found by replacing Eq. (11) by the close-to-it equation

$$f = \lambda P_\tau f + (1 - \lambda)h, \quad 0 < \lambda < 1, \tag{12}$$

where h is an arbitrary function satisfying conditions

$$\|h\|_{L_1} = 1, \quad h \geq 0. \tag{13}$$

It is easy to see that for $0 < \lambda < 1$ the operator in the right-hand side of Eq. (12) is strictly contractive. For every $0 < \lambda < 1$ there exists exactly one solution f_λ. The following theorem holds true:

Theorem 2. If τ satisfies the assumptions of Theorem 1, then for every h satisfying conditions (13) there exists a strong limit

$$\lim_{\lambda \to 1} f_\lambda = f$$

that satisfies Eq. (11), and consequently f is the density of the invariant measure.

Equation (12) has an interesting probabilistic interpretation. The operator $\lambda P_\tau + (1 - \lambda)h$ corresponds to the following process.* Given x_n we draw lots between two possibilities, (I) and (II). The first of these possibilities has probability λ, the other one $(1 - \lambda)$. If (I) is drawn, we take $x_{n+1} = \tau(x_n)$. If (II) is drawn, we choose the point x_{n+1} by lot (independently from x_n) according to the probability distribution for which the function h is the density. It is intuitively obvious that if λ is close to 1, our new process is almost deterministic. Almost always $x_{n+1} = \tau(x_n)$. Another method, convenient for numerical calculations, of seeking an invariant measure is by replacing the operator P_τ by a close-to-it stochastic matrix. We take

$$p_{ij}^n = n \int_{\Delta_{jn}} P_\tau 1_{\Delta_{in}}(x)\, dx \quad \text{where} \quad \Delta_{in} = \left[\frac{i-1}{n} \right]. \tag{14}†$$

Denote by $u_1{}^n, \ldots, u_n{}^n$ a vector satisfying the equation

$$u_j^n = \sum_{i=1}^n p_{ij}^n u_i^n \tag{15}$$

and conditions

$$u_i^n \geq 0, \quad \sum_{i=1}^n u_i^k = 1. \tag{16}$$

---

* Speaking formally, we should consider the operator

$$\lambda P_\tau f + (1 - \lambda)h \int_0^1 f(s)\, ds.$$

For the functions f under consideration we can neglect the last factor as equal to one.
  † 1_Δ denotes the characteristic function of the set Δ.

The matrix (p_{ij}^n) is stochastic $(p_{ij}^n \geq 0, \sum_{j=1}^n p_{ij}^n = 1)$, so such a vector exists. Ulam anticipated that when $n \to \infty$, the sequence of functions

$$f^n(x) = n \sum_{i=1}^n u_i^n 1_{\Delta_{ni}}(x) \qquad (17)$$

converges to the density of measure invariant under τ. The positive answer to this problem is given by the following theorem proved by Li and Yorke [7].

Theorem 3. If the assumptions of Theorem 1 are satisfied, then the sequence f^n given by formula (17) is compact in L_1 and converges to the set of solutions of Eq. (11), i.e.,

$$\lim_n \inf_f \| f^n - f \|_{L_1} = 0, \qquad (18)$$

where the infimum is taken over all the solutions of Eq. (11).

In particular, it follows that the limit of every convergent subsequence $\{f^n\}$ is a solution of Eq. (11), so it is the density of an invariant measure. If the (absolutely continuous) invariant measure is unique, then f^n converges to the density of this measure.

5. Applications

Our first example deals with the rotary drilling of rocks. The drilling tool (in the simplest case) has the form of a cog-wheel that rolls along the surface of the rock. If the rolling velocity is high enough, then after every blow of a tooth on the rock surface, the tool rebounds before the next blow. The energy of a blow, and consequently the efficiency of the tool, depends on the angle at which the blow occurred. Denote this angle, at the nth strike, by x_n, and assume that it can change within the interval $[0, T]$. For a normalized sequence $s_n = x_n/T$ we can write (see [4])

$$s_{n+1} = \tau_F(s_n), \qquad (19)$$

where $\tau_F: [0, 1] \to [0, 1]$ is a rather complicated function, namely,

$$\tau_F(s) = \text{frac}\{s + \alpha q(s) - [(\alpha q(s))^2 + 2\alpha s q(s) - \alpha q(s)(1 + q(s))]^{1/2}\}. \qquad (20)$$

In this formula

$$\alpha = \frac{F}{F-1}, \qquad q(s) = 1 - \text{int}\left(\frac{1-2s}{\alpha-1}\right),$$

$$\text{int}(x) = \begin{cases} 0, & \text{for } x < 1 \\ \text{the largest integer less than or equal to } x, & \text{for } x \geq 1. \end{cases}$$

$$\text{frac}(x) = x - \text{int}(x).$$

The specific form of the function τ_F is not relevant for our purposes. The fact of basic importance is simply that it depends on the single constant F, which has a simple physical interpretation, being Freude's number:

$$F = Mu^2/QR.$$

Here M is the mass of the tool, R its radius, Q the pressure, and u the linear velocity of rolling. It is easy to see that as the value of F increases, the incline of the graph of function τ_F gets steeper. We can show that

$$\inf\left| d\tau_F^2(s)/ds \right| > 1 \tag{21}$$

for $F > 2$. So, for $F > 2$, according to Theorem 1, there exists an absolutely continuous invariant measure, which we denote μ_F.

As we said before, knowing the angle of the blow $x_n = Ts_n$, we can find the energy of the blow. It is also known that the volume of rock broken off by each blow is proportional to the energy of the blow. Finally, therefore, the volume V of rock broken off is the function of the angle of blow s, i.e., $V = V(s)$. Due to the individual ergodic theorem we have

$$\overline{V} = \lim_n \frac{1}{n} \sum_{k=1}^n V(s_k) = \int_0^1 V(s)\mu_F\,(ds). \tag{22}$$

So, knowing the invariant measure, we are able to count the mean efficiency per blow. In a similar way, it is possible to find a mean expenditure per blow. The received formulas appear to be in accordance with drilling practice (see [4]). They can also be useful for optimizing drilling tool efficiency.

Our second example is connected with the technology of foundry casting. After cooling, the molds must be cleaned of sand, clay, and other materials sticking to them. One of the most commonly used methods is shaking. The simplest model of such a process can be imagined in the following way: The item to be cleaned is lying on a movable horizontal surface. Appropriately placed machinery sets the surface into vertical periodic vibration. As a result of these vibrations, the mold is tossed up, and then hits the surface again. As practice shows, a large amount of energy is absorbed during such an operation. Assuming that the collisions are ideally nonelastic, we again obtain a one-dimensional process. Let us assume that the period of vibrations of the working surface is equal to T, and let us denote by t_n the moment in which the nth blow of the mold being cleaned occurs. Knowing the quantity $s_n = t_n/T$ it is possible to obtain s_{n+1}. It turns out that the function obtained is similar to the function given by formula (20), and depends on the parameter F, which is Freude's number

$$F = A/T^2 g,$$

where A is the amplitude of vibrations and g the gravity constant. In contrast to our first example, here condition (7) is not satisfied, even for large values of F. But assuming that the process is partially stochasticly perturbed (for instance, because of some pauses in the work of the machinery), instead of the measure μ_F corresponding to the deterministic process, we can use the measure $\mu_{F\lambda}$ whose density satisfies Eq. (12) for $\tau = \tau_F$, $h \equiv 1$. It appears that, calculated with the use of $\mu_{F\lambda}$, the integrals

$$\int_0^1 \varphi(x)\mu_{F\lambda}(dx) \tag{23}$$

do not depend in any way that matters on λ when λ is close to 1. From the practical point of view, this is the solution of the problem, because by the use of the integrals of form (23) it is possible to estimate all quantities of interest to an engineer such as the efficiency of the process, the mean use of the working surface, etc. It also indicates that Theorems 1 and 2 can occur by much more general assumptions, and furthermore, the integrals (23) may have limits with $\lambda \to 1$, even if the limit measure is not absolutely continuous.

6. Final Remarks

Theorem 1 belongs to the class of theorems in which the existence of an invariant measure is connected with the feature of "expanding" described by assumption (7). In the case of the n-dimensional manifolds for expanding diffeomorphisms, the existence of a nonbanal invariant measure was proved by Krzyżewski and Szlenk [3] and Avez [1]. For the mappings of the set

$[0, 1]^n$ into itself, some generalizations of Rényi were obtained by Waterman [10].

In Banach spaces the situation is more difficult. It can be shown that for every infinite-dimensional Banach space there exists a continuous mapping of a closed unit ball into itself for which an invariant measure (defined on a σ-algebra of Borel sets) does not exist. On the other hand, the existence of invariant measures for mappings in Banach spaces seems to be very important. Because, as Foiaş [2] has shown, a good, mathematically correct description of turbulent motion given by the Navier–Stokes equations can be obtained by considering the invariant measures in the space of initial states of the velocity field.

Finally, let us note that the effect of turbulence can be easily noticed in the system described by Eqs. (19) and (20). For Freude's number $F < 2$ there exists exactly one periodic stable solution. For $F > 2$ stable solutions do not exist, and the sequence $\{s_n\}$ is dense in support of measure μ_F.

REFERENCES

[1] A. Avez, Propriétés ergodiques des endomorphismes dilatants compactes, *C. R. Acad. Sci. Paris, Sér. A-B* **266** (1968), A610–A612. MR 37 No. 6944.

[2] C. Foiaş, Statistical study of Navier–Stokes equation II, *Rend. Sem. Mat. Univ. Padova* **49** (1973), 9–123.

[3] K. Krzyżewski and W. Szlenk, On invariant measures for expanding differentiable mappings, *Stud. Math.* **33** (1969), 83-92. MR 39 No. 7067.

[4] A. Lasota and P. Rusek, An application of ergodic theory to the determination of the efficiency of cogged drilling bits (Polish with Russian and English summaries). *Arch. Górnictwa* **19** (1974), 281–295.

[5] A. Lasota and J. A. Yorke, On the existence of invariant measures for piecewise monotonic transformations, *Trans. Amer. Math. Soc.* **186** (1973), 481-488.

[6] T. Y. Li and J. A. Yorke, Ergodic transformations from an interval into itself, *Trans. Amer. Math. Soc.* (to be published).

[7] T. Y. Li and J. A. Yorke, A numerical method for calculating absolutely continuous invariant measures (to be published).

[8] A. Rényi, Representation for real numbers and their ergodic properties, *Acta Math. Acad. Sci. Hungary* **9** (1957), 477-493. MR 20 No. 3843.

[9] S. M. Ulam, "A Collection of Mathematical Problems" (Interscience Tracts in Pure and Appl. Math., No. 8). Wiley (Interscience), New York, 1960. MR 22 No. 10884.

[10] M. S. Waterman, Some ergodic properties of multidimensional F-expansions, *Z. Wahrscheinlichkeitstheorie Verw. Gebiete* **16** (1970), 77-103. MR 44 No. 173.

[11] A. A. Kosjakin, and E. A. Sandler, Ergodic properties of a certain class of piecewise smooth transformations of a segment, *Izv. Vysš. Učebn. Zaved. Matematika* **118** (1972), 32-40. MR 45 No. 8802.

Bifurcation Theory for Odd Potential Operators*

R. R. HUILGOL†
Department of Mechanics and Mechanical Aerospace Engineering
Illinois Institute of Technology, Chicago, Illinois

1. Existence of Eigenvalues and Eigenfunctions

Let A and B be odd mappings from a real infinite-dimensional Banach space X into its dual X^*. Consider the *eigenvalue problem* for the pair (A, B), namely, the problem of finding $u \in X$, an *eigenfunction* satisfying some normalization conditions, and a real number λ, an *eigenvalue*, such that

$$A(u) = \lambda B(u). \tag{1.1}$$

Let A and B be potential operators with potentials a and b, respectively, such that

Assumption I. $a(u) = 0 \Leftrightarrow u = 0$, $\quad b(u) = 0 \Leftrightarrow u = 0$, $\quad$ and $\quad a(u) > 0$, $b(u) > 0$ for all $u \neq 0$.

Following Amann [1], we demand that A and B obey the following:

Assumption II. $A: X \to X^*$ is an odd potential operator that is uniformly continuous on bounded sets and satisfies condition $(S)_1$. For a given constant $r > 0$, the level set $M_r(a) \equiv \{u \in X \,|\, a(u) = r\}$ is bounded and each ray through the origin intersects $M_r(a)$. Moreover, for every $u \neq 0$, $\langle A(u), u \rangle > 0$ and there exists a constant $\rho_r > 0$ such that $\langle A(u), u \rangle \geq \rho_r$ on $M_r(a)$.

Assumption III. $B: X \to X^*$ is a strongly sequentially continuous odd potential operator, i.e., B maps every weakly convergent sequence into a strongly convergent sequence, and $b(u) \neq 0 \Rightarrow B(u) \neq 0$.

We now record the following

* This research was supported by NSF Grant GP-31312.

† Present address: School of Mathematical Sciences, The Flinders University of South Australia, Bedford Park, Australia.

Theorem 1 (Amann [1]). Let X be an infinite-dimensional uniformly convex Banach space. Let the operators A and B obey Assumptions (II) and (III). Then the eigenvalue problem (1.1) has infinitely many distinct eigenfunctions obeying the normalization condition $a(u) = r$ provided

$$\gamma\{u \in M_r(a)|b(u) \neq 0\} = \infty. \tag{1.2}$$

Moreover, let k be a positive integer and let $\mathscr{C}_k(r) \equiv \{C \subset M_r(a)|C$ symmetric, compact, gen$(C) \geq k\}$, and define

$$\beta_k(r) = \sup_{C \in \mathscr{C}_k(r)} \inf_{u \in C} b(u). \tag{1.3}$$

Then if $\beta_k > 0$, there exists an eigenfunction $u_k \in M_r(a)$ of (1.1) with

$$b(u_k) = \beta_k(r). \tag{1.4}$$

For a definition of $(S)_1$ and $\gamma\{\cdot|\cdot\}$, see [1].

Remarks. (1) Amann's theorem guarantees the existence of infinitely many eigenvalues and eigenfunctions for the problem (1.1) under conditions less severe than those used in [2], which was based on Theorem 1 in [3].

(2) Since $\beta_k(r) > 0$ for any $r > 0$ because of Assumption I, for all $r > 0$, there exists at least one eigenfunction $u_k \in M_r(a)$ in the problem under study here.

(3) $M_r(a)$ is homeomorphic to the unit sphere in X.

As a preliminary to the next section, we lay down the following

Definition. The eigenvalue λ_k of a problem such as (1.1) is given by

$$1/\lambda_k \equiv \beta_k(1) = \sup_{C \in \mathscr{C}_k(1)} \inf_{u \in C} b(u). \tag{1.5}$$

2. Bifurcation from an Eigenvalue

Let it now be assumed that the operator B is the sum of B_1 and a higher-order term B_2, i.e.,

$$B = B_1 + B_2, \tag{2.1}$$

with a precise meaning to be given to this "higher-order" term in (2.9). Let b_1 and b_2 be the even functionals associated with B_1 and B_2, respectively. Let $b = b_1 + b_2$, and b_1 both obey Assumptions I and III. We impose some additional restrictions on a and b_1 next.

UPPER AND LOWER BOUNDS

Let there exist f_1, f_2, g_1, g_2 such that

$$f_1(t)a(u) \leq a(tu) \leq f_2(t)a(u),$$
$$g_1(t)b_1(u) \leq b_1(tu) \leq g_2(t)b_1(u), \qquad (2.2)$$

where f_1, f_2, g_1, g_2: $[0, \infty) \to [0, \infty)$ are homeomorphisms onto such that

(i) $f_i(0) = g_i(0) = 0$, $i = 1, 2$;

(ii) $\lim_{r \to 0} g_1(f_2^{-1}(r))/g_2(f_1^{-1}(r)) = 1$. $\qquad (2.3)$

In addition, f_i and g_i are extended to the whole real line by assuming them to be even, i.e.,

$$f_i(-t) = f_i(t), \qquad g_i(-t) = g_i(t), \qquad i = 1, 2. \qquad (2.4)$$

Let $u \in M_1(a)$. Then $\beta_k^{(1)}(1)$ is given by [with $\lambda_k^{(1)}$ the corresponding eigenvalue for the problem $a(u) = \lambda b_1(u)$]:

$$1/\lambda_k^{(1)} \equiv \beta_k^{(1)}(1) = \sup_{C \in \mathscr{C}_k(1)} \inf_{u \in C} b_1(u). \qquad (2.5)$$

Let $tu = v$, $-\infty < t < \infty$, and let $v \in M_r(a)$. Then $f_2^{-1}(r) \leq t \leq f_1^{-1}(r)$. Thus

$$\beta_k^{(1)}(r) = \sup_{C \in \mathscr{C}_k(r)} \inf_{v \in C} b_1(v) \leq g_2(f_1^{-1}(r))/\lambda_k^{(1)}. \qquad (2.6)$$

Similarly, $\beta_k^{(1)}(r) \geq g_1(f_2^{-1}(r))/\lambda_k^{(1)}$. Next,

$$\sup_{C \in \mathscr{C}_k(r)} \inf_{u \in C} b_1(u) + \inf_{u \in M_r(a)} b_2(u) \leq \sup_{C \in \mathscr{C}_k(r)} \inf_{u \in C} (b_1(u) + b_2(u))$$

$$\leq \sup_{C \in \mathscr{C}_k(r)} \inf_{u \in C} b_1(u) + \sup_{u \in M_r(a)} b_2(u), \qquad (2.7)$$

i.e.,

$$\frac{g_1(f_2^{-1}(r))}{\lambda_k^{(1)}} + \inf_{u \in M_r(a)} b_2(u) \leq \beta_k(r) \leq \frac{g_2(f_1^{-1}(r))}{\lambda_k^{(1)}} + \sup_{u \in M_r(a)} b_2(u). \qquad (2.8)$$

We shall now make the meaning of "higher order" precise by demanding that b_2 obey the condition:

HIGHER ORDER OF b_2

Along each ray from the origin to $u \in M_r(a)$,

$$\lim_{r \to 0} \frac{\sup_{u \in M_r(a)} |b_2(u)|}{g_2(f_1^{-1}(r))} = 0. \tag{2.9}$$

Using (2.3) and (2.9) in (2.8), we obtain

$$\lim_{r \to 0} \frac{\beta_k(r)}{g_2(f_1^{-1}(r))} = \frac{1}{\lambda_k^{(1)}}. \tag{2.10}$$

Now, for a given $\beta_k(r)$, by Theorem 1, there exists a $u_k \in M_r(a)$ such that

$$\beta_k(r) = b(u_k). \tag{2.11}$$

Clearly, as $r \to 0$, $u_k \to 0$. Since (2.10) can be rewritten as

$$\lim_{r \to 0} \frac{\beta_k(r)}{r} \frac{r}{g_2(f_1^{-1}(r))} = \frac{1}{\lambda_k^{(1)}}, \tag{2.12}$$

we have proved

Theorem 2. Let the operators A and B obey Assumptions II and III; let the potentials a and b obey Assumption I, (2.2)–(2.4), and (2.9). Let $\lambda_k^{(1)}$ be a nonzero eigenvalue of the problem

$$A(u) = \lambda B_1(u). \tag{2.13}$$

Then $\lambda_k^{(1)}$ is a bifurcation point of the eigenvalue problem

$$A(u) = \lambda(B_1(u) + B_2(u)), \tag{2.14}$$

provided the limit

$$\lim_{r \to 0} \frac{g_2(f_1^{-1}(r))}{r} = 1 \tag{2.15}$$

holds.

The above theorem generalizes the result of Krasnosel'skii [4, p. 332], who assumed that $A = I$, the identity operator, and $B = L + N$, where L is linear and N nonlinear; our theorem generalizes the case treated more recently by Fučik et al. [2], who assumed that a and b_1 are homogeneous operators of the same degree. In addition, our proof, while based on that given in [2], demands that b_2 be of "higher order" along rays only.

REFERENCES

[1] H. Amann, Lusternik–Schnirelman theory and nonlinear eigenvalue problems, *Math. Ann.* **199** (1972), 55-72.

[2] S. Fučik, J. Nečas, J. Souček, and V. Souček, Krasnoselskii's main bifurcation theorem, *Arch. Rational Mech. Anal* **54** (1974), 328–339.

[3] S. Fučik and J. Nečas, Lusternik–Schnirelmann theorem and nonlinear eigenvalue problems, *Math. Nachr.* **53** (1972), 277-289.

[4] M. A. Krasnosel'skii, "Topological Methods in the Theory of Nonlinear Integral Equations." Pergamon, Oxford, 1964.

An Existence Theorem for Solutions of Orientor Fields

CZESŁAW OLECH

Institute of Mathematics
Polish Academy of Sciences, Warsaw, Poland

Introduction

By an orientor field we mean a relation

$$\dot{x} \in F(t, x), \tag{1}$$

where F is a map from $[0, 1] \times R^n$ into compact subsets of R^n. An absolutely continuous function x from $[0, 1] = I$ into R^n is a solution of (1) if $\dot{x}(t) \in F(t, x(t))$ almost everywhere (a.e.) in I.

The aim of this paper is to present the following theorem concerning existence of a solution of (1) satisfying initial condition

$$x(0) = a. \tag{2}$$

Theorem. Assume that:

(i) F is measurable in t for each fixed x;
(ii) F is upper semicontinuous in x for each fixed t;
(iii) For each x and $y \in F(t, x)$ the norm $|y| \leq \lambda(t)$, where λ is integrable on I;
(iv) If for a fixed (t_0, x_0), $F(t_0, x_0)$ is not convex, then $F(t_0, x)$ is continuous in x with respect to the Hausdorff distance of compact subsets of R^n at the point $x = x_0$.

Under those assumptions for each $a \in R^n$ there is a solution of (1) defined on I and satisfying (2).

There are two special cases of the theorem that we would like to mention. One is when $F(t, x)$ is assumed to be convex everywhere. In this case, assumption (iv) is superfluous and the existence of a solution of (1) has been proved by many authors (cf., for example, Ważewski [7] or Filippov [2]). On the other hand, if F is upper semicontinuous only but not convex, then there may not exist a solution of (1). For example, $F(t, x) = \{1\}$ if $x > 0$, $\{-1\}$ if $x < 0$, and $\{1, -1\}$ if $x = 0$ is upper semicontinuous but there does not exist a solution of (1) satisfying the initial condition $x(0) = 0$. Hermes [4] has posed the question: Does there exist a solution of (1) if instead of (iii), (iv) is assumed for each x and t? This question has been answered affirmatively first by Filippov [3] when F is assumed to be continuous in both variables, and later extended by Kaczyński and Olech [5] to the Caratheodory type of assumptions; that is, (i), (ii), and (iv) for each x and any t fixed. This is the second special case of our theorem. Also, Antosiewicz and Cellina [1] obtained the same result as a corollary to a theorem they proved, concerning existence of continuous selection to a set-valued map induced by F from the space of continuous functions into subsets of L_1 space.

The detailed proof of the theorem is to be found in [6]. It follows very much the proof of the existence theorem given in [5]. Here we restrict ourselves only to some indication of the difficulties one has to face if the convexity assumption is not present, and we show how a sequence of approximate solutions is constructed.

1. Approximate Solutions

The existence of solutions of both ordinary differential equations and orientor fields is usually obtained by constructing a sequence $x_n(t)$ of approximate solutions, which contains a uniformly convergent subsequence, and then the limit function is proved to be the solution sought. Such a sequence has the property

$$d(\dot{x}_n(t), F(t, x_n(t) + \varepsilon_n(t))) \to 0 \qquad \text{as} \quad n \to \infty, \tag{3}$$

where d stands for the distance of a point from a set and $\varepsilon_n(t) \to 0$ uniformly on I. If $x_n(t) \to x_0(t)$ uniformly and $|\dot{x}_n(t)| \le \lambda(t)$, then $\dot{x}_n \to \dot{x}_0$ weakly in L_1 and (3) implies

$$\dot{x}_0(t) \in \bigcap_{\varepsilon > 0} \text{cl co} \bigcup_{|x - x_0(t)| < \varepsilon} F(t, x) \qquad \text{a.e. in } I, \tag{4}$$

where cl co stands for closed convex hull. If (ii) holds then the right-hand side of (4) is equal to co $F(t, x_0(t))$; thus if $F(t, x)$ is assumed to be convex (4) implies that $x_0(t)$ is a solution of (1). Thus if convexity of F is assumed, then any x_n satisfying (3) leads to the existence of a solution of (1).

However, if we do not assume convexity then we need to know more about convergence of the derivatives of the approximate solutions and because of (ii) it is enough to know that $\dot{x}_n(t) \to \dot{x}_0(t)$ a.e. in $[0, 1]$. This would be the case if $\{\dot{x}_n\}$ is precompact in the strong topology of L_1. Hence in this case the construction of approximate solutions has to be finer.

2. Construction of Approximate Solutions

Let $\{r_i\}$ be a decreasing sequence of reals tending to zero, and for each i, let A_i be a finite subset of the ball B centered at a and of radius $r = \int_0^1 \lambda(t)\, dt$, such that for each $x \in B$ there is $a_1 \in A_i$ such that $|x - a_i| < r_i/2$.

Let $a_i \in A_i$, $i = 1, \ldots, k$, be such that $|a_i - a_{i-1}| \le r_i$, $i = 2, \ldots, k$. For each such sequence we define an integrable function $u_{a_1, \ldots, a_k}(t)$ such that

$$u_{a_1, \ldots, a_k}(t) \in F(t, a_k) \qquad \text{a.e. in } I, \tag{5}$$

and

$$\left| u_{a_1, \ldots, a_k}(t) - u_{a_1, \ldots, a_{k-1}}(t) \right| \le h(F(t, a_k), F(t, a_{k-1})), \tag{6}$$

where h stands for the Hausdorff distance between two sets. Existence of such functions follows from the assumption and an induction argument.

Let $\{h_i\}$ be a decreasing sequence of reals such that $1/h_i$ and h_i/h_{i+1} are integers and $\int_t^{t+h_i} \lambda(t)\, dt < r_i/4$ for each $t \in I$.

The nth approximate solution is obtained from

$$x_n(0) = a \qquad \text{and} \qquad \dot{x}_n(t) = u_{a_1{}^n(t), \ldots, a_n{}^n(t)}(t), \tag{7}$$

where $a_i{}^n(t) \in A_i$ and is constant on intervals $[kh_i, (k+1)h_i)$,

$$\left| a_i{}^n(t) - a_{i-1}^n(t) \right| \le r_i$$

and

$$\left| x_n(kh_i) - a_i{}^n(kh_i) \right| \le r_i/2, \qquad m = 0, \ldots, 1/h_i - 1, \quad i = 1, \ldots, n. \tag{8}$$

By an induction argument one can show that (7) and (8) can be satisfied. Note that the derivative of x_n, n fixed, is chosen from a finite set of functions. From (6) and (7) we can estimate the difference $\dot{x}_{n+p}(t) - \dot{x}_n(t)$, and if F is continuous in x and r_i are sufficiently small this estimate can be made small in L_1 norm and uniform with respect to p. Thus $\dot{x}_n$ would be precompact in L_1 topology. From (8) and the definition of h_i it follows that x_n satisfies (3). Thus without any loss of generality we can assume that $x_n(t) \to x_0(t)$ uniformly in I. Hence by (4) $\dot{x}_0(t) \in \text{co } F(t, x_0(t))$. Thus one only has to show that $\dot{x}_0(t) \in F(t, x_0(t))$ a.e. in T, where $T = \{t \,|\, F(t, x_0(t)) \text{ is}$

not convex}. This is obtained by proving that for each $s > 0$ $\{\dot{x}_n(t)\}$ is L_1-precompact on T_s, where $T_s = \{t \mid F(t, x) \text{ is not convex if } |x - x_0(t)| \le s\}$. To get the latter condition (iv) is essentially used.

REFERENCES

[1] H. A. Antosiewicz and A. Cellina, Continuous selections and differential relations, *J. Differential Equations* (to appear).

[2] A. F. Filippov, Differential equations with multivalued right-hand side (in Russian), *Dokl. Akad. Nauk SSSR* **151** (1963), 65–68.

[3] A. F. Filippov, On existence of solution of multivalued differential equations (in Russian), *Mat. Zametki* **10** (1971), 307-313.

[4] H. Hermes, The generalized differential equation $\dot{x} \in R(t, x)$, *Advan. Math.* **4** (1970), 149-169.

[5] H. Kaczyński and C. Olech, Existence of solutions of orientor fields with nonconvex right-hand side, *Ann. Polon. Math.* **29** (1974), 61–66.

[6] C. Olech, Existence of solutions of nonconvex orientor fields *Boll. Un. Mat. Ital.* (to appear).

[7] T. Ważewski, Systèmes de commande et équations au contingent, *Bull. Acad. Polon. Sci., Ser. Sci. Math. Astronom. Phys.* **9** (1961), 139-160.

Nonlinear Perturbations at Resonance

R. KANNAN

Department of Mathematics
Michigan State University, East Lansing, Michigan

In this paper we consider nonlinear differential equations of the type $Lu + \lambda u = Nu$, where L is a linear self-adjoint differential operator over a real Hilbert space S with preassigned linear homogeneous boundary conditions, λ an eigenvalue of the associated linear problem $Lu + \lambda u = 0$, and N a nonlinear operator over S. In [7] Landesman and Lazer considered the nonlinear problem

$$Lu + \lambda u + g(u) = h \quad \text{in} \quad D, \qquad u = 0 \quad \text{on} \quad \partial D,$$

where D is a smooth bounded domain in R^n and g a real-valued continuous function such that the limits $\lim_{s \to \infty} g(s) = g(\infty)$ and $\lim_{s \to -\infty} g(s) = g(-\infty)$ are finite, and further $g(-\infty) \le g(s) \le g(+\infty)$ for all s. They obtained sufficient conditions for the existence of weak solutions to the above nonlinear problem. Their work was a generalization of earlier work of Lazer and Leach [8] in ordinary differential equations involving bounded perturbations of forced harmonic oscillators at resonance. The results of Landesman and Lazer have been further extended and improved by Nečas [10], Williams [12], Mawhin [9], and others.

1

Here we present sufficient conditions for the existence of solutions of the abstract semilinear problem $Lu + \lambda u = Nu$ under suitable hypotheses on N. These sufficient conditions are then studied in the light of some of the results of the previously mentioned authors. In the second part of this paper we extend these ideas to nonlinear problems where the nonlinearity is not defined over the entire Hilbert space S. We consider in this section those differential operators L that together with the boundary conditions admit of a natural decomposition in the form TT^*.

We apply the ideas developed in [3] to split the nonlinear differential equation into an equivalent system of two Hammerstein equations and study them for the existence of solutions by using some of the recent results of nonlinear functional analysis. It must be mentioned that this idea of splitting, motivated by the work of Cesari [1], is the underlying idea in all of the papers mentioned above.

2

Let L be a linear self-adjoint differential operator with preassigned homogeneous boundary conditions over a smooth bounded domain in R^n and let S be the real Hilbert space $L_2(\Omega)$ with norm and inner product denoted by $\| \cdot \|$, $\langle , \rangle$, respectively. Further, let L be such that the associated eigenvalue problem $Lu + \lambda u = 0$ has a countable system of real eigenvalues $\{\lambda_i\}, 0 \le \lambda_1 \le \lambda_2 \le \cdots, \lambda_i \to +\infty$. Also, let the corresponding eigenfunctions $\{\phi_i\}$ form a complete orthonormal system in S. We consider the nonlinear problem

$$Lu + \lambda_m u = N_1 u, \tag{1}$$

where $N_1 : S \to S$ is a nonlinear operator.

Let $S_0 = \{\phi_1, \ldots, \phi_m\}$ and let $P: S \to S_0$ be the projection operator. Also let $S = S_0 \oplus S_1$. Let $H: S_1 \to S_1$ be the linear operator such that

(a) $H(I - P)Lu = (I - P)u, u \in \mathscr{D}(L)$,
(b) $PLu = LPu, u \in \mathscr{D}(L)$, and
(c) $LH(I - P)Nu = (I - P)Nu, u \in S$, and $Nu = N_1 u - \lambda_m u$.

Then the nonlinear problem (1) is equivalent to the system of equations (for details see [1])

$$u - H(I - P)Nu = Pu, \tag{2}$$

$$PNu - PLu = 0. \tag{3}$$

Let u^* be any arbitrary element of S_0. If $u \in S$ is a solution of the equation

$$u - H(I - P)Nu = u^*, \tag{4}$$

then clearly $Pu = u^*$ and hence u is a solution of (2). Thus, if Eq. (2) has a solution $u \in S$ for each $u^* \in S_0$, the system of Eqs. (2) and (3) is equivalent to the system of equations

$$u - H(I - P)Nu = u^*, \tag{4}$$

$$PN[I - H(I - P)N]^{-1}u^* - Lu^* \ni 0. \tag{5}$$

We now state the following theorem.

Theorem 1. Let $N_1: S \to S$ be such that

(i) N_1 is continuous and bounded, i.e., N_1 takes bounded sets into bounded sets;

(ii) There exists $p < \lambda_{m+1} - \lambda_m$ such that for all $u, v \in S$

$$\langle N_1 u - N_1 v, u - v \rangle \geq - p \|u - v\|^2;$$

(iii) There exists $R > 0$ such that for all $u^* \in S_0$ satisfying $Pu = u^*$,

$$\langle N_1 u, u^* \rangle \leq 0,$$

where $u = [I - H(I - P)N]^{-1}u^*$.

Then the nonlinear problem (1) has at least one solution.

Remark 1. We have made the assumption $0 \leq \lambda_1$. However, the arguments involved in proving Theorem 1 can be easily modified to consider the case when a finite number of the λ_i's are negative (see Cesari [2]).

Remark 2. Hypothesis (iii) of Theorem 1 is implied by the sufficiency hypotheses of Landesman and Lazer [7], Nečas [10], and others. In [10] Nečas considers the nonlinear problem

$$Lu = Nu + h,$$

where $h \in S$ and

(i) $N: S \to S$ is completely continuous,
(ii) $\|Nu\| \leq \alpha < \infty$ for all $u \in S$, and
(iii) 0 is an eigenvalue of the associated linear problem.

Nečas proves the following: Let the limit $\lim_{t \to \infty} < N(u + tu^*), u^* > = l(u^*)$ exist and be finite, where the limit is uniform with respect to u on bounded sets of S and with respect to all $u^* \in S_0 = \mathcal{N}(L)$ with $\|u^*\| = 1$. If $u^* \in \mathcal{N}(L) = S_0$ and $\|u^*\| = 1$ implies

$$\langle h, u^* \rangle + l(u^*) < 0, \tag{6}$$

then the nonlinear problem $Lu = Nu + h$ has at least one solution.

It can be proved that hypothesis (iii) of Theorem 1 is implied by hypothesis (6) of Nečas. Similar remarks can be made about the results of Landesman and Lazer [7] and Schatzman [11]. It must be pointed out, however, that hypothesis (ii) does not hold in these papers and thus it would be interesting to extend Theorem 1 to these cases.

3

In this section we point out how by using the natural decomposition of linear differential operators L in the form TT^*, one can extend the ideas of the earlier section to large nonlinearities that are not defined over all of S. Operators of the type TT^* have been studied by Kato [6] and others. Simple examples of operators L admitting a decomposition of the type TT^* are

(i) $Lu = -u''$, $u(0) = u(2\pi)$, $u'(0) = u'(2\pi)$,
(ii) $Lu = \Delta^2 u$, $u = \partial u/\partial n = 0$ on Ω.

The natural decomposition of L in the form TT^* induces a decomposition of $-H(I - P)$ in the form J^*J and thus (4) reduces to

$$u + J^*JNu = u^*$$

or

$$w + J^*JN(w + u^*) = 0, \qquad \text{where} \quad w = u - u^*.$$

Using the fact that J^* is one-to-one, the above equation reduces to $v + JN(J^*v + u^*) = 0$. With this modified form of the auxiliary equation we now have the following theorem.

Theorem 2. Let $N: \mathscr{D}(N) = \mathscr{D}(T^*) \to S$ be a nonlinear operator such that

(i) There exists $p \geq 0$ with $p < \lambda_{m+1} - \lambda_m$ such that
$$\langle Nu - Nv, u - v \rangle \geq -p\|u - v\|^2 \qquad \text{for all} \quad u, v \in S;$$

(ii) N is continuous from $\mathscr{D}(T^*)$ to S;
(iii) There exists a function $\gamma: R^+ \to R^+$ such that $u \in \mathscr{D}(N)$ with $|u|_{\mathscr{D}(N)} \leq R$ implies
$$\|Nu\| \leq \gamma(R);$$

(iv) There exists $R > 0$ such that for all $u^* \in S_0$ with $\|u^*\| = R$, $\langle Nu, u^* \rangle \leq 0$, where $u = [I + J^*JN]^{-1}u^*$.

Then the nonlinear problem

$$Lu + \lambda_m u = Nu$$

has at least one solution.

Thus Theorem 2 essentially extends Theorem 1 to the case where N is not necessarily defined over S.

The proof of Theorem 2 may be seen in Kannan and Locker [5], where an extensive study is made of the nature of $L = TT^*$ and the properties of its eigenfunctions in relation to those of $L_1 = T^*T$, and these properties are utilized to consider existence of solutions of nonlinear boundary value problems. Similar ideas are also applied in Dunninger and Locker [4] to the case where L is the biharmonic operator.

REFERENCES

[1] L. Cesari, Functional analysis and Galerkin's method, *Michigan Math. J.* **11** (1964), 385–414.

[2] L. Cesari, *Proc. Con. Differential Equations, Los Angeles, 1974.*

[3] L. Cesari and R. Kannan, Functional analysis and nonlinear differential equations, *Bull. Amer. Math. Soc.* **79** (1973), 1216–1219.

[4] D. R. Dunninger and J. Locker, Monotone operators and nonlinear biharmonic boundary value problems, *Pacific J. Math.* (to appear).

[5] R. Kannan and J. Locker, Nonlinear boundary value problems and operators TT^* (to appear).

[6] T. Kato, On some approximate methods concerning the operators T^*T, *Math. Ann.* **126** (1953), 253–262.

[7] E. M. Landesman and A. C. Lazer, Nonlinear perturbation of elliptic boundary value problems at resonance, *Indiana Univ. Math. J.* **19** (1970), 609–623.

[8] A. C. Lazer and D. E. Leach, Bounded perturbations of forced harmonic oscillators at resonance, *Ann. Mat. Pura Appl.* **82** (1969) 49–68.

[9] J. Mawhin, this volume.

[10] J. Nečas, On the range of nonlinear operators with linear asymptotes which are not invertible, *Commun. Math. Univ. Carolinae* **14** (1973), 63–72.

[11] M. Schatzman, Probléms aux limites nonlinéaires semicoercifs, *C. R. Acad. Sci. Paris* **215** (1972), 1305–1308.

[12] S. A. Williams, A sharp sufficient condition for solution of a nonlinear elliptic boundary value problem, *J. Differential Equations* **8** (1970), 580–586.

On Continuous Dependence of Fixed Points of Condensing Maps

ZVI ARTSTEIN*
Lefschetz Center for Dynamical Systems
Division of Applied Mathematics
Brown University, Providence, Rhode Island

We shall present a result on continuity with respect to the parameter of the fixed points of the operators

$$T_\lambda \colon X \to X.$$

Hale [4] gave sufficient conditions for the continuous dependence when the operators are α-condensing. We shall show in what sense the conditions are necessary conditions and cannot be improved. A more general theory that applies to other systems will appear in [2].

We assume X is a Banach space, the parameter λ belongs to a metric space Λ, and the family T_λ ($\lambda \in \Lambda$) is collectively α-condensing. Let us give the definition of the last statement. The Kuratowski measure α of non-compactness associates with a subset B of X the number $\alpha(B) = \inf\{r : B$ can be covered by a finite number of sets with diameter less than $r\}$. An operator T is α-condensing if for every $B \subset X$ the inequality $\alpha(TB) \geq \alpha(B)$ implies $\alpha(B) = 0$. The family T_λ ($\lambda \in \Lambda$) is collectively α-condensing if for every $B \subset X$, $\alpha(\bigcup_{\lambda \in \Lambda} T_\lambda B) \leq \alpha(B)$ and equality implies $\alpha(B) = 0$. Notice that implicitly we assume that the range is bounded.

For background, examples, and applications of α-condensing and α-contraction operators see Hale [3, 4]. Many differential and integral equations as well as functional differential equations can be reduced to a fixed-point equation $x = Tx$, where T is α-condensing.

We do not want to assume existence or uniqueness of solutions. Therefore, the mapping $s(\lambda)$ that associates with each λ the set of solutions of the equation $x = T_\lambda x$ is a *multivalued* mapping (with the possibility of empty values). As a continuity concept we choose the *upper semicontinuity*. [A multivalued function $s(\lambda)$ is upper semicontinuous if for each open set Q of the range space the set $\{\lambda : s(\lambda) \subset Q\}$ is an open set.] This

* The research was supported in part by the Office of Naval Research under NONR N000-14-67-A-0191-000906, and in part by the National Science Foundation under GP 28931X2.

concept is natural not only because we are able to prove upper semi-continuity, but the important fact is that upper semicontinuity implies continuity in the case of uniqueness. Also if $s(\lambda)$ is upper semicontinuous and has compact values, then $\lim \lambda_n = \lambda$ and $x_n \in s(\lambda_n)$ imply the existence of an element $x \in s(\lambda)$ and a subsequence x_m such that $\lim x_m = x$. In particular, if $s(\lambda) = \{x\}$ then $\lim x_n = x$.

The following result was proved by Hale [4, Theorem 1], although in a slightly different terminology. Hale also shows that the theorem includes and simplifies many former results.

Theorem 1. $s(\lambda)$ is upper semicontinuous if $T_\lambda x$ is continuous in λ and x simultaneously.

The natural question now arises: Is the joint continuity also a necessary condition? Can we add the "only if" to the theorem? The answer is negative and there is a good reason for this. The only values of T_λ that are involved in "creating" $s(\lambda)$ are the fixed points of the operator. Without any restrictions one could change T_λ on the complement of $s(\lambda)$ and maintain the same set of fixed points. Therefore, $s(\lambda)$ might be even constant without any restrictions on $T_\lambda x$ for $x \notin s(\lambda)$. So a global condition on $T_\lambda x$ cannot be necessary. Still something can be said. Let us add a forcing term to the equation, i.e., consider the equation

$$x = T_\lambda x + y.$$

Such a forcing term appears naturally in many integral equations. Let $s(\lambda, y)$ be the set of solutions of $x = T_\lambda x + y$. We shall see below that Hale's conditions are sufficient also for the upper semicontinuity of $s(\lambda, y)$ in (λ, y). Now, when y "travels" along X, all the values of T_λ get involved in solving the equation, and there is hope of getting necessary conditions. Indeed, Hale's conditions are now necessary. In other words, if the (quite reasonable) requirement is added, namely, continuity also with respect to the forcing term, then the conditions become necessary.

Theorem 2. $s(\lambda, y)$ is upper semicontinuous if and only if $T_\lambda x$ is continuous in λ and x simultaneously.

We shall give a compressed proof. We write T_n instead of T_{λ_n}. The "if" part is a slight modification of [4, Theorem 1]. Let $\lambda_n \to \lambda$ and $y_n \to y$. Let $x_n \in s(\lambda_n, y_n)$, i.e., $x_n = T_n x_n + y_n$.

Then

$$\alpha(\{x_n\}) = \alpha(\{T_n x_n + y_n\}) \le \alpha(\{T_n x_n\}) + \alpha(\{y_n\}) \le \alpha\left(\bigcup_m T_m\{x_n\}\right).$$

The first inequality follows from the properties $\alpha(A + B) \le \alpha(A) + \alpha(B)$ and $A \subset B$ implies $\alpha(A) \le \alpha(B)$. Since $\{y_n\}$ is precompact it follows that $\alpha(\{y_n\}) = 0$ and now the second inequality follows from the inclusion $\{T_n x_n\} \subset \bigcup_m T_m\{x_n\}$. Thus

$$\alpha(\{x_n\}) \le \alpha\left(\bigcup_m T_m\{x_n\}\right)$$

which together with the condensity assumption implies $\alpha(\{x_n\}) = 0$, i.e., $\{x_n\}$ is a precompact set. The joint continuity of $T_\lambda x$ in (λ, x) implies that each limit point of $\{x_n\}$ is a solution of $x = T_\lambda x + y$, and the upper semi-continuity follows. For the "only if" part assume the contrary that $\lambda_n \to \lambda_0$, $x_n \to x$, but $\|T_n x_n - T_0 x\| \ge \varepsilon > 0$ for $n = 1, 2, \ldots$. Define y_n by $y_n = x_n - T_n x_n$. Since $\{x_n\}$ is precompact and T_λ ($\lambda \in \Lambda$) is collectively α-condensing, it follows that $\{T_n x_n\}$ and hence $\{y_n\}$ is precompact. Suppose that y_k ($k = 1, 2, \ldots$) is a converging subsequence with limit y. Now we have $(\lambda_k, y_k) \to (\lambda_0, y)$, $x_k \to x$, and $x_k \in s(\lambda_k, y_k)$, and therefore the upper semi-continuity implies $x \in s(\lambda_0, y)$, i.e., $x = T_0 x + y$. Thus $T_k x_k = y_k - x_k$ converges to $T_0 x = y - x$, a contradiction. This completes the proof.

An open problem. I do not know whether Theorem 2 still holds when the parameter set Λ is a general topological space (and not a metric space). This generalization is true if α-contraction is considered instead of α-condensing.

REFERENCES

[1] Z. Artstein, Continuous dependence of solutions of Volterra integral equations, *SIAM J. Math. Anal.* **6** (1975), 446–456.
[2] Z. Artstein, Continuous dependence of solutions of operator equations, Lefschetz Center for Dynamical Systems, Brown Univ., preprint.
[3] J. K. Hale, α-contraction and differential equations, *Proc. Equations Différentiel Fon. Nonlin., Brussels 1973*, pp. 15-42. Herman, Paris, 1973.
[4] J. K. Hale, Continuous dependence of fixed points of condensing maps, *J. Math. Anal.* **46** (1974), 388-399.

Small Noise Ergodic Dynamical Systems

CHARLES HOLLAND

Department of Mathematics
Purdue University, West Lafayette, Indiana

We outline some recent results in small noise problems in the ergodic case and indicate some possible implications for small noise ergodic control problems. Suppose that the state of the dynamical system evolves according to the stochastic differential equations

$$d\xi_x^\varepsilon(t) = f(\xi_x^\varepsilon(t))\, dt + (2\varepsilon)^{1/2} B\, dw(t), \qquad \xi_x^\varepsilon(0) = x, \tag{1}$$

where w is brownian motion of appropriate dimension. Let us assume that for each $\varepsilon > 0$ the process (1) generates a unique ergodic measure μ_ε, i.e.,

$$\lim_{t \to \infty} \mathrm{Prob}[\xi_x^\varepsilon(t) \in B] = \mu^\varepsilon(B),$$

for all borel subsets B of R^n and for any initial condition x. We are interested in establishing conditions guaranteeing that

$$\int L(x)\, d\mu_\varepsilon(x) = L(0) + \sum_{k=1}^{n} \varepsilon^k b_k + o(\varepsilon^k), \tag{2}$$

for constants b_k, where $o(\varepsilon)\varepsilon^{-n} \to 0$ as $\varepsilon \to 0$. In [1] the following theorem was established.

Theorem. Let the following hold:

(i) $f(x) = Ax + Bg(x)$, g is C^∞ on R^n, $g(0) = 0$, g and its partial derivatives of all orders are bounded.

(ii) (A, B) is controllable and satisfies (CO).

(iii) There exists a matrix $Q > 0$ and a constant $c > 0$ such that for all $x \in R^n$, $f_x'(x)Q + Q f_x(x) < -cI$.

(iv) L is C^∞, and L and its derivatives of all orders are of polynomial growth.

Then (2) is true for any positive integer n.

The assumptions of the theorem guarantee the existence of a unique ergodic measure μ_ε. This has been shown in [3] where condition (CO) is defined. Assumption (iii) implies exponential asymptotic stability of the corresponding deterministic system with $\varepsilon = 0$. It would be interesting to see if this assumption could be weakened to $f_x{'}(0)Q + Qf_x(0) < -CI$ and global asymptotic stability of the deterministic system. An example in [1] shows that (iii) cannot be replaced by global asymptotic stability of the deterministic system alone.

The constants b_k can be found by solving linear algebraic equations with coefficients involving the partial derivatives of f and L evaluated at the origin. The details of this procedure are discussed in [2], where we considered the two-dimensional system

$$d\xi_1 = \xi_2 \, dt, \qquad d\xi_2 = -4\xi_2 - 2\xi_1 - \sin \xi_1 \, dt + (2\varepsilon)^{1/2} \, dw(t), \qquad (3)$$

which arises from the formal equation

$$\ddot{\xi} + 4\dot{\xi} + 2\xi + \sin \xi = (2\varepsilon)^{1/2}\dot{w}(t). \qquad (4)$$

Simple calculations show that

$$\int x_1{}^2 \, d\mu_\varepsilon(x) = \frac{\varepsilon}{12} + \frac{\varepsilon^2}{144} + o(\varepsilon^2), \qquad (5)$$

and

$$\int x_1{}^4 \, d\mu_\varepsilon(x) = \frac{\varepsilon^2}{48} + o(\varepsilon^2). \qquad (6)$$

For this example the density of the ergodic measure can be calculated explicitly. The quantities on the left-hand sides of (5) and (6) can then be approximated by a numerical integration procedure.

Ergodic control problems. Let $Y^0(x)$ be the optimal feedback control in the infinite-time deterministic control problem. Suppose that with use of Y^0 the state equations and cost function for the control problem are such that the theorem is satisfied. Then use of Y^0 in the corresponding ergodic control problem (minimization of the cost function with respect to the resulting ergodic measure) yields a cost of $d\varepsilon + o(\varepsilon)$ for some constant d. Under certain assumptions one could expect that the optimal cost would also satisfy $d\varepsilon + o(\varepsilon)$. Then Y^0 would be a reasonably good suboptimal control in the ergodic problem for sufficiently small ε. In those cases one would like to have an effective method of computing Y^0 or some approximation to Y^0 that yields the expansion $d\varepsilon + o(\varepsilon)$.

Wonham and Cashman [5] used statistical linearization as a technique for determining a suboptimal control. Recently, Lasry [4] has studied ergodic problems under assumptions that included periodicity of the state equations and cost function in the state variables. The above approach is not valid for these problems since the deterministic system with use of Y^0 will not be globally asymptotically stable.

REFERENCES

[1] C. Holland, Ergodic expansions in small noise problems, *J. Differential Equations* **16** (1974), 281–288.

[2] C. Holland, Stationary small noise problems, *Internat. J. Non-Linear Mech.* **10** (1975).

[3] H. J. Kushner. The Cauchy problem for a class of degenerate parabolic equations and asymptotic properties of the related diffusion processes, *J. Differential Equations* **6** (1969), 209–231.

[4] J. Lasry, Controle stationary asymptotique, Mathematiques de la decision, Univ. de Paris, IX (1974).

[5] W. M. Wonham and W. F. Cashman, A computational approach to optimal control of stochastic saturating systems, *Internat. J. Control* **10** (1969), 77-98.

Chapter 3 : EVOLUTIONARY EQUATIONS

"Pointwise Degeneracy" for Delay Evolutionary Equations

PIERRE CHARRIER

U.E.R. de Mathématiques et Informatique
Université de Bordeaux, Talence, France

I. Introduction

A well-known property of linear differential systems in $\mathbb{R}^n$ is that the set of values of all solutions at a given time t is $\mathbb{R}^n$ itself. This property does not extend to delay differential systems. There exist systems with lag for which the set of values of all solutions is a proper subspace of $\mathbb{R}^n$ (for some time t). They are called "pointwise degenerate." They were studied first by Popov [1], and then by other authors, for instance, Asner and Halanay [2], Charrier and Haugazeau [3], Choudhury [4], and Kappel [5].

Our goal here is to look at this property in infinite-dimensional spaces (see Charrier [6]). H will be a separable Hilbert space; we denote by $(\, , \,)_H$ the scalar product. $T(t)$ is a bounded, strongly continuous semigroup on H and A is its infinitesimal generator, with domain $D(A)$. For some important examples (see Example 1) the "attainable set" $\mathfrak{A}(t) = \{T(t)y_0 \,|\, y_0 \in D(A)\}$ is dense in H. In any case, is it possible to find a perturbation $By(t - h)$ such that for the delay equation

$$\frac{d}{dt} y(t) = Ay(t) + By(t - h),$$

$$y(0) = y_0; \qquad y(\theta) = \Phi(0), \qquad \theta \in [-h, 0], \tag{1}$$

the "attainable set" is a proper subset of H?

II. Transformation of the Problem. A Necessary and Sufficient Condition

We choose to work with compact operators B. The initial data Φ are assumed to be continuously differentiable from $[-h, 0[$ into H. Then we can prove (see Kato [7]) existence and uniqueness of a solution for (1).

Definition 1. Equation (1) is said to be "pointwise degenerate" at time t_1 with respect to vectors q_j of H $(j = 1, 2, \ldots, p)$ iff every solution of (1) satisfies

$$(q_j, y(t_1))_H = 0, \qquad j = 1, 2, \ldots, p.$$

Remark. As in finite dimensions it is easy to see that if an equation is "pointwise degenerate" at time t_1, it remains "pointwise degenerate" at every time $t \geq t_1$. Because of the continuity of y, the set of degeneracy is a closed interval $[t_0, +\infty[$. For simplicity we study systems for which degeneracy occurs at time $2h$. Let us define

$$Y(t) = \begin{bmatrix} y(t) \\ y(t+h) \end{bmatrix}, \qquad t \in [0, h]. \tag{2}$$

Following the method proposed in Charrier and Haugazeau [3] one can prove that to every solution y of (1) there corresponds a solution of

$$dY(t)/dt = \mathscr{A}Y(t) + \mathscr{B}\Psi(t), \tag{3}$$

$$Y(0) = \begin{bmatrix} y(0) \\ y(h) \end{bmatrix}, \qquad Y(h) = \begin{bmatrix} y(h) \\ y(2h) \end{bmatrix}, \tag{4}$$

and conversely, where $\mathscr{A}$ is the infinitesimal generator of a semigroup $\mathscr{T}(t)$ bounded, strongly continuous on $H \times H$, and represented by the matrix of operators

$$\mathscr{A} = \begin{bmatrix} A & 0 \\ B & A \end{bmatrix},$$

$\mathscr{B}$ is a compact operator on $H \times H$ defined by

$$\mathscr{B} = \begin{bmatrix} B \\ 0 \end{bmatrix},$$

and $\psi(t) = \Phi(t - h)$.

Therefore, Eq. (1) is "pointwise degenerate" with respect to the vectors q_j at time $2h$ iff any solution of (3) and (4) satisfies

$$\left(\begin{bmatrix} 0 \\ q_j \end{bmatrix}, Y(h) \right)_{H \times H} = 0.$$

In Eq. (3), ψ occurs as a control. We obtain a condition of degeneracy as a consequence of the controllability properties of this infinite-dimensional control system. We shall denote by $\mathscr{R}(h)$ the reachable set of the control system (3) starting from the null initial condition. Equation (3) is said to be completely controllable iff $\overline{\mathscr{R}(h)} = H \times H$. If $\mathscr{R}(h)$ is not dense we shall call

$\mathscr{R}(h)^\perp$ the supplementary space of $\mathscr{R}(h)$ in $H \times H$ and P (respectively, P') the orthogonal projection on $\mathscr{R}(h)$ [resp. $\mathscr{R}(h)^\perp$].

Lemma 1. Problem (3), (4) has a solution only if $(y(0), y(h), y(2h)) \in H^3$ belongs to the kernel of application $L : H^3 \to \mathscr{R}(h)$,

$$(y(0), y(h), y(2h)) \to P' \begin{bmatrix} y(h) \\ y(2h) \end{bmatrix} - P' \mathscr{T}(h) \begin{bmatrix} y(0) \\ y(h) \end{bmatrix}.$$

Now let us define the application G:

$$H^3 \to \mathbb{R}^p, \quad (y(0), y(h), y(2h)) \to \begin{bmatrix} (q_1, y(2h))_H \\ \vdots \\ (q_p, y(2h))_H \end{bmatrix}.$$

Proposition 1. Equation (1) is "pointwise degenerate" with respect to vectors q_j at time $2h$ iff (i) $P' \neq 0$, (ii) $\ker L \subset \ker G$.

To prove this proposition we use Lemma 1 and the definition of Y as in Charrier and Haugazeau [3, Theorem 2].

This NASC can be stated without difficulty at time kh (for any positive integer k), but it is clear that such a condition cannot lead us to an algebraic condition as in finite dimensions. So we need sufficient conditions that will be useful for further applications.

III. Sufficient Conditions. Example

Proposition 2. Suppose that q_j belongs to $D(A^*), j = 1, 2, \ldots, p$. If there exists a compact operator Z from H into H with range in $D(A)$ such that

(i) $ZB = 0$ and $B = AZ - ZA$,
(ii) $T(h)^*q_j = Z^*q_j, j = 1, 2, \ldots, p$, and $Z^*T(h)^*q_j = 0, j = 1, 2, \ldots, p$.

Then Eq. (1) is "pointwise "degenerate" with respect to vectors q_j at time $2h$.

The idea of the proof is that the uncontrollable part of (3) is isomorphic to H and satisfies the equation

$$\frac{dy}{dt} w(t) = Aw(t) \quad \text{with} \quad w(t) = y(t) + Zy(t - h).$$

Corollary 1. Let q_j be elements of $D(A^*)$. Let us suppose that there exist vectors r_j, $j = 1, \ldots, p$, in $D(A)$ such that

$$(q_j, r_k)_H = \delta_{j,k}, \qquad (q_j, T(h)r_k)_H = 0, \qquad \text{and} \qquad (q_j, T(h)Ar_k)_H = 0. \qquad (5)$$

Then if B is defined by

$$Bv = \sum_{j=1}^{p} (T(h)^*q_j, v)_H Ar_j - \sum_{j=1}^{p} (T(h)^*A^*q_j, v)_H r_j,$$

Eq. (1) is "pointwise degenerate" with respect to the vectors q_j at time $2h$.

Example 1. $H = L^2(]0, 1[)$, $D(A) = \{v \in H_0^1(]0, 1[) \cap H^2(]0, 1[)\}$ $A = d^2/dx^2$. Then Eq. (1) is the equation of the heat diffusion on a rod with Dirichlet conditions.

The set $\mathfrak{A}(t) = \{T(t)y_0 \,|\, y_0 \in D(A)\}$ is dense in H for any $t \geq 0$. A is a self-adjoint operator. The set of eigenfunctions w_k of A is orthonormal complete in H. Let us define $q_j = \sum_1^\infty \alpha_k^j w_k$ in $D(A)$. The $r_j = \sum_1^\infty \gamma_k^j w_k$, $j = 1, \ldots, p$, can be chosen to satisfy (5) if $3p$ coefficients α_k^j are not null. From the corollary, every solution of

$$\frac{\partial y}{\partial t}(t, x) = \frac{\partial^2 y}{\partial x^2} + \int_0^1 k(x, \xi)y(t - h, \xi)\, d\xi,$$

$$\text{with} \quad k(x, \xi) = \sum_{j=1}^{p} ([T(h)q_j](\xi)r_j''(x) - [T(h)q_j''](\xi)r_j(x)),$$

$$y(t, 1) = y(t, 0) = 0,$$
$$y(0, x) = y_0(x), \qquad y(t, x) = \Phi(t, x), \quad t \in [-h, 0[,$$

satisfies $\int_0^1 q_j(x)y(t, x)\, dx = 0$ for $t \geq 2h$. Moreover, in this example $T(t)$ is a holomorphic semigroup and we can take q in H, for instance, $q(x) \equiv 1$. Then any solution has a null mean on $[0, 1]$ for $t \geq 2h$.

Remark. If the semigroup $T(t)$ is holomorphic (see Kato [7]) we can prove that the minimum time of degeneracy is kh, where k is an integer greater than or equal to 2. Similar work can be done for systems with two lags and for systems including a lag in a derivative.

Proposition 3. q_j is in $D(A^*)$, r_j in $D(A)$, $j = 1, 2, \ldots, p$, and satisfy

$$(q_k, r_j)_H = \delta_{kj}, \qquad (q_k, T(h)r_j)_H = 0, \qquad j = 1, \ldots, p.$$

Let us define

$$B_1 v = \sum_{j=1}^{p} (T(h)^* q_j, v)_H \, Ar_j - (T(h)^* A^* q_j, v)_H r_j,$$

$$B_2 v = \sum_{j,k=1}^{p} (T(h)^* A^* q_j, r_k)_H (T(h)^* q_k, v)_H r_j,$$

$$B_3 v = \sum_{j=1}^{p} (T(h)^* q_j, v)_H \, Ar_j,$$

$$B_4 v = - \sum_{j=1}^{p} (T(h)^* q_j, v)_H r_j.$$

Then the equations $dy/dt = Ay(t) + B_1 y(t - h) + B_2 y(t - 2h)$ and $dy(t)/dt = Ay(t) + B_3 y(t - h) + B_4 \, dy(t - h)/dt$ are degenerate with respect to vector q_j at time $2h$ for the first and at time h for the second.

Remark.   If we work with vectors q_j in H, the operators B are in general no longer compact. With the assumptions of Example 1 it is possible to extend Corollary 1 to an infinite enumerable (but not dense) family of q_j (for example, $q_k = w_{2k} + w_{2k-1}$, $k = 1, 2, \ldots$).

IV. A Consequence of "Pointwise Degeneracy"

Let z_m be an orthonormal complete family of H. We denote by $L(t)z_m$ the solution of

$$dy(t)/dt = A^* y(t) + B^* y(t - h), \qquad y(0) = z_m, \quad \Phi \equiv 0. \tag{6}$$

Lemma 2.   The solution of

$$dy(t)/dt = Ay(t) + By(t - h), \qquad y(0) = y_0, \quad \Phi \equiv 0 \tag{7}$$

is given by $y(t) = L(t)^* y_0 = \sum_m (L(t)z_m, y_0)z_m$.

Proposition 4.   Consider the control system $dy(t)/dt = Ay(t) + u(t)$, $y(0) = y_0$; let us suppose that y_0 and r satisfy $(y_0, r)_H = 1$, $(y_0, T(h)^* r)_H = (y_0, T(h)^* A^* r)_H = 0$, and that B is defined by $Bv = (A^* r, v)_H \, T(y)y_0 - (r, v)_H \, T(h)Ay_0$. Then if we take for control $u(t) = 0$ on $[0, h[$, $u(t) = By(t - h)$, $t \geq h$, the solution is stabilized to the origin at time $2h$ [i.e., $y(t) = 0$ for $t \geq 2h$]. Moreover, if $T(t)$ is holomorphic $2h$ is the minimum time (for control of this given form).

We remark that B depends on y_0. If we take $Bv = \sum_{j=1}^{p} (A^*r_j, v)_H$ $\times T(h)y_0^j - (r_j, v)_H T(h)Ay_0^j$, then B depends only on the subspace V spanned by the y_0^j and not on y_0 taken in V. In this case B appears as a feedback associated to the subspace V.

REFERENCES

[1] Popov, V. M., Pointwise degeneracy of linear time invariant delay differential equations, *J. Differential Equations* **11** (1972), 541–561.
[2] Asner, B. A., and Halanay, A. Algebraic theory of pointwise degenerate delay differential systems, *J. Differential Equations* **14** (1973), 293–306.
[3] Charrier, P., and Haugazeau, Y., On the degeneracy of linear time invariant delay differential systems, *J. Math. Anal. Appl.* **52** (1975), 42–55.
[4] Choudhury, A. K., Pointwise degeneracy of second order linear time invariant delay differential systems with distributed lag, private communication.
[5] Kappel, F., Degeneracy of functional differential equations, private communication.
[6] Charrier, P., Sur la dégénérescence des équations d'évolution avec retard. *C. R. Acad. Sci. Paris* **277** (1973), 121–123.
[7] Kato, T., "Perturbation Theory for Linear Operator." Springer-Verlag, Berlin, 1966.

On Constructing a Liapunov Functional While Defining a Linear Dynamical System*

J. A. WALKER

Department of Mechanical Engineering
Northwestern University, Evanston, Illinois
and
Lefschetz Center for Dynamical Systems
Division of Applied Mathematics
Brown University, Providence, Rhode Island

We consider the problem of applying Liapunov's direct method to an autonomous physical system, initially described by only a formal equation of motion. Our problem involves not only construction of a Liapunov functional, but also the selection of an appropriate state space $\mathcal{X}$ on which the given formal equation can be associated with an abstract dynamical system [1]. Assuming that the system is linear and that some appropriate (but unknown) state space is a Hilbert space, we suggest a systematic operational approach to the entire problem.

Restating our objectives, we wish to obtain a Hilbert space $\mathcal{X}$ and a densely defined linear operator $A: \mathcal{D}(A) \to \mathcal{X}$ such that the abstract evolution equation

$$\frac{dx(t)}{dt} = Ax(t), \qquad t \geq 0,$$

$$x(0) = \phi \in \mathcal{D}(A),$$

(1)

is an appropriate abstraction of the given formal equation, and such that A generates a linear C_0-semigroup $\{T(t)\}_{t \geq 0}$ on $\mathcal{X}$ [2]. Moreover, we wish to find a continuous functional $V: \mathcal{X} \to \mathcal{R}$ such that $\dot{V}(x) \leq 0$ for all $x \in \mathcal{X}$, where

$$\dot{V}(x) \equiv \limsup_{t \downarrow 0} \frac{1}{t} [T(t)x - x].$$

If $\mathcal{D}(A)$ is dense in the Hilbert space $\mathcal{X}$, then a sufficient condition for A to be a generator is that $A - \omega I$ be maximal dissipative for some real

---

* This research was supported by the National Science Foundation under GK 40009.

number ω [3]. Although this condition is not necessary, it is known that if $\{\hat{T}(t)\}_{t \geq 0}$ is a linear C_0-semigroup on a Hilbert space $\hat{\mathscr{X}}$, then there exists a (possibly larger) Hilbert space $\mathscr{X}$ and an extension $\{T(t)\}_{t \geq 0}$ of $\{\hat{T}(t)\}_{t \geq 0}$ such that $\{T(t)\}_{t \geq 0}$ is a C_0-semigroup on $\mathscr{X}$ with generator A; moreover, $A - \omega I$ is maximal dissipative for some ω [4, 5]. Hence, for our problem, there must exist some appropriate state space such that $A - \omega I$ is maximal dissipative for some ω, where A is the generator on $\mathscr{X}$. We also note that if $A - \omega I$ is maximal dissipative for some $\omega \leq 0$, then $T(t)$ is a contraction and $V(x) \equiv \langle x, x \rangle_{\mathscr{X}}$ is a Liapunov functional.

The suggested approach to our problem utilizes an artifice. Referring to the given formal equation of motion, the first step is to choose (somewhat arbitrarily) a convenient Hilbert space $\mathscr{H}$ and a densely defined linear operator $\tilde{A} : (\mathscr{D}(\tilde{A}) \subset \mathscr{H}) \to \mathscr{H}$, such that the formal equation seems closely related to the abstract equation

$$\frac{dy(t)}{dt} = \tilde{A}y(t), \qquad t \geq 0. \tag{2}$$

However, $\tilde{A}$ need not be a generator; in fact, (2) need not even have nontrivial solutions. Define a Banach space $\mathscr{Y}$ as the completion of $\mathscr{D}(\tilde{A})$ in $\|\cdot\|_{\mathscr{Y}}$, where $\|y\|_{\mathscr{Y}} \equiv \|y\|_{\mathscr{H}} + \|\tilde{A}y\|_{\mathscr{H}}$ for $y \in \mathscr{D}(\tilde{A})$, and note that the mapping $\tilde{A} : (\mathscr{D}(\tilde{A}) \subset \mathscr{Y}) \to \mathscr{H}$ is bounded. An abstract operational approach to our problem is now suggested by the following result:

Theorem 1. Let $\mathscr{D}(\tilde{A}) = \mathscr{Y}$ and let there exist real numbers μ, $\omega < \mu$, $\gamma > 0$, such that

(a) $\mu I - \tilde{A} : \mathscr{Y} \to \mathscr{H}$ is a homeomorphism;
(b) For some bounded linear operator $\tilde{B} : \mathscr{Y} \to \mathscr{H}$,
      (i) $\langle \tilde{B}y_1, y_2 \rangle_{\mathscr{H}} = \langle \tilde{B}y_2, y_1 \rangle_{\mathscr{H}}, \forall y_1, y_2 \in \mathscr{Y}$,
      (ii) $\langle \tilde{B}y, y \rangle_{\mathscr{H}} \geq \gamma \langle y, y \rangle_{\mathscr{H}}, \forall y \in \mathscr{Y}$,
      (iii) $\mathrm{Re}\langle \tilde{B}y, (\tilde{A} - \omega I)y \rangle_{\mathscr{H}} \leq 0, \forall y \in \mathscr{Y}$.
Define a Hilbert space $\mathscr{X}$ as the completion of $\mathscr{Y}$ in $\|\cdot\|_{\mathscr{X}}$, where $\langle y_1, y_2 \rangle_{\mathscr{X}} \equiv \langle By_1, y_2 \rangle_{\mathscr{H}}$ for $y_1, y_2 \in \mathscr{Y}$. Define $A : (\mathscr{D}(A) \subset \mathscr{X}) \to \mathscr{X}$ to be the restriction of $\tilde{A}$ to $\mathscr{D}(A) \equiv \{y \in \mathscr{Y} \mid \tilde{A}y \in \mathscr{X}\}$. Then A generates a linear C_0-semigroup on $\mathscr{X}$.

The proof is given in [5]. Noting that $\mathscr{D}(A) = (\mu I - \tilde{A})^{-1}(\mathscr{X})$, it follows that $\mathscr{D}(A)$ is dense in $\mathscr{X}$. Then it need only be shown that $\mathscr{R}(\mu I - A) = \mathscr{X}$ and $A - \omega I$ is dissipative; hence, $A - \omega I$ is maximal dissipative [3].

By using Theorem 1 we obtain an appropriate state space and an abstract equation (1); moreover, if $\omega \leq 0$, $T(t)$ is a contraction and $V(x) \equiv \langle x, x \rangle_{\mathscr{X}}$ is a Liapunov functional for (1). By complicating the definition of A, the conditions of Theorem 1 can be weakened. The following result is obtained in [5].

Theorem 2. Let there exist real numbers μ, $\omega < \mu$, $\alpha > 0$, such that

(a) $\mathscr{R}(\mu I - \tilde{A})$ is dense in $\mathscr{H}$ and

$$\|y\|_{\mathscr{Y}} \leq \alpha \|(\mu I - \tilde{A})y\|_{\mathscr{H}}, \qquad \forall y \in \mathscr{D}(\tilde{A}),$$

(b) For some bounded linear operator $\tilde{B}: \mathscr{Y} \to \mathscr{H}$,

    (i) $\langle \tilde{B}y_1, y_2 \rangle_{\mathscr{H}} = \overline{\langle \tilde{B}y_2, y_1 \rangle}_{\mathscr{H}}, \forall y_1, y_2 \in \mathscr{Y}$,

    (ii) $\langle \tilde{B}y, y \rangle_{\mathscr{H}} \geq 0, \forall y \in \mathscr{Y}$,

    (iii) $\operatorname{Re}\langle \tilde{B}y, (\tilde{A} - \omega I)y \rangle_{\mathscr{H}} \leq 0, \forall y \in \mathscr{D}(\tilde{A})$.

Define a Hilbert space $\mathscr{X}$ as the completion of $\mathscr{Y}$ in $\|\cdot\|_{\mathscr{X}}$, where $\langle y_1, y_2 \rangle_{\mathscr{X}} \equiv \langle \tilde{B}y_1, y_2 \rangle_{\mathscr{H}}$ for $y_1, y_2 \in \mathscr{Y}$. Let $A_1: \mathscr{Y} \to \mathscr{H}$ be the continuous extension of $\tilde{A}: (\mathscr{D}(\tilde{A}) \subset \mathscr{Y}) \to \mathscr{H}$, and let $A_2: (\mathscr{D}(A_2) \subset \mathscr{X}) \to \mathscr{X}$ be the restriction of A_1 to $\mathscr{D}(A_2) \equiv \{y \in \mathscr{Y} | A_1 y \in \mathscr{X}\}$. Then A_2 has a unique extension $A: (\mathscr{D}(A) \subset \mathscr{X}) \to \mathscr{X}$ such that $A - \omega I$ is maximal dissipative; moreover, A generates a linear C_0-semigroup on $\mathscr{X}$.

In applications it is more important that A be consistent with the formal equation than that it be rigorously derived from the artificial operator $\tilde{A}$. Hence, given a formal equation, it may be both simpler and more general merely to define $\tilde{A}: \mathscr{D}(\tilde{A}) \to \mathscr{H}$, $\mathscr{D}(\tilde{A})$ dense in $\mathscr{H}$, and define a normed linear space $\mathscr{Y}$ by

$$\|y\|_{\mathscr{Y}} = \sum_{p=0}^{n} \|\tilde{A}^p y\|_{\mathscr{H}}, \qquad y \in \mathscr{D}(\tilde{A}^n),$$

for some positive integer n. We then seek $\tilde{B}$ satisfying (b) of Theorem 2, define $\mathscr{X}$ as in Theorem 2, and refer directly to the formal equation in an attempt to define $A: \mathscr{D}(A) \to \mathscr{X}$ such that $\mathscr{D}(A)$ is dense in $\mathscr{X}$ and $A - \omega I$ is maximal dissipative. If this attempt is successful, A generates $\{T(t)\}_{t \geq 0}$; moreover, if $\omega \leq 0$, $T(t)$ is a contraction and $V(x) \equiv \langle x, x \rangle_{\mathscr{X}}$ is a Liapunov functional. Several extensions and applications of this approach are described in [5].

REFERENCES

[1] J. K. Hale, Dynamical systems and stability, *J. Math. Anal. Appl.* **26** (1969), 39–59.
[2] A. Friedman, "Partial Differential Equations." Holt, New York, 1969.
[3] R. S. Phillips, Dissipative operators and hyperbolic systems of partial differential equations, *Trans. Amer. Math. Soc.* **90** (1959), 679–698.
[4] M. Slemrod, An application of maximal dissipative sets in control theory, *J. Math. Anal. Appl.* **46** (1974), 369–387.
[5] J. A. Walker, On the application of Liapunov's direct method to linear dynamical systems, CDS Tech. Rep. 74-7, Div. of Appl. Math., Brown Univ., Providence, Rhode Island (to appear in *J. Math. Anal. Appl.*).

Measurability and Continuity Conditions for Evolutionary Processes

JOHN M BALL*
Lefschetz Center for Dynamical Systems
Division of Applied Mathematics
Brown University, Providence, Rhode Island

1. Introduction

Let X be a topological space. By definition, an *evolutionary process* on X is a family of operators $U(t, s): X \to X$, defined for $t \in \mathbb{R}^+$, $s \in \mathbb{R}$, and satisfying (i) $U(0, s) = $ identity; (ii) $U(t + \tau, s) = U(t, s + \tau)U(\tau, s)$ for t, $\tau \in \mathbb{R}^+$, $s \in \mathbb{R}$. Such processes arise in the mathematical modelling of nonautonomous systems, when $U(t, s)x$ represents the position (or state) at time $t + s$ of the point that at time s was at x. In the special case when the operators $U(t, s) \overset{\text{def}}{=} T(t)$ are independent of s, the evolutionary process defines a *semigroup* $\{T(t)\}$, $t \in \mathbb{R}^+$.

In the paper [1] it was shown that in certain situations measurability and continuity properties known to be satisfied for a semigroup could be strengthened using the semigroup properties. These results directly generalize to nonlinear semigroups those known for semigroups of continuous linear operators on a Banach space. We extend this work to evolutionary processes.

2. Results for Evolutionary Processes

Throughout this section $\{U(t, s)\}$ denotes an evolutionary process on X satisfying the hypothesis:

(A)   For each $t \in \mathbb{R}^+$ the map $(s, x) \mapsto U(t, s)x$ is (jointly) sequentially continuous from $\mathbb{R} \times X \to X$.

---

* Present address: Department of Mathematics, Heriot-Watt University, Edinburgh, Scotland.

Theorem 1. Let X be a metric space. Suppose that for each $s \in \mathbb{R}$, $x \in X$, the map $t \mapsto U(t, s)x$ is strongly (Lebesgue) measurable on $(0, \infty)$. Then the map $(t, s, x) \mapsto U(t, s)x$ is continuous on $(0, \infty) \times \mathbb{R} \times X$.

Theorem 2. Let X be arbitrary. Suppose that for each $s \in \mathbb{R}$, $x \in X$, the map $t \mapsto U(t, s)x$ is Baire continuous on $(0, \infty)$ and, when restricted to the complement of some first category set, has second countable range. Then the map $(t, s, x) \mapsto U(t, s)x$ is sequentially continuous on $(0, \infty) \times \mathbb{R} \times X$.

Theorem 3. Let X be a subset of a Banach space. Suppose that for each $s \in \mathbb{R}$, $x \in X$, the map $t \mapsto U(t, s)x$ is weakly continuous from the right on $(0, \infty)$. Then the map $(t, s, x) \mapsto U(t, s)x$ is continuous on $(0, \infty) \times \mathbb{R} \times X$ with respect to the norm topology on X.

Theorem 4. Let X be a subset of a uniformly convex Banach space. Suppose that

(a) For each $s_1, s_2 \in \mathbb{R}$, $x_1, x_2 \in X$, $t_n \to 0_+$ implies
$$\liminf_{n \to \infty} \| U(t_n, s_1)x_1 - U(t_n, s_2)x_2 \| \le \| x_1 - x_2 \|;$$

(b) For each $s \in \mathbb{R}$, $x \in X$, the map $t \mapsto U(t, s)x$ is weakly continuous from the right at $t = 0$.

Then for each $s \in \mathbb{R}$, $x \in X$, the map $t \mapsto U(t, s)x$ is continuous on $[0, \infty)$ with respect to the norm topology on X.

The proofs of Theorems 1–4 may be found in [2]. The method is to prove the results first in the semigroup case, and then to apply the semigroup theorems to the semigroup $\{S(t)\}$, $t \in \mathbb{R}^+$, which is defined on the space $\mathbb{R} \times X$ by

$$S(t)\binom{s}{x} \overset{\text{def}}{=} \binom{s+t}{U(t, s)x}. \tag{1}$$

To show that (A) is satisfied it is sufficient to prove that $U(t, s)x = u(t, f(s), x)$ for some functions u and f, where $f : \mathbb{R} \to Y$ is continuous, Y is a topological space, and $u(t, \cdot, \cdot) : Y \times X \to X$ is sequentially continuous for each $t \in \mathbb{R}^+$. In applications $f(s)$ may be the s-translate of a function

$g: \mathbb{R} \to Y_1$, where Y_1 is a topological space, $f(s)(t) \stackrel{\text{def}}{=} g(s + t)$ for $t \in \mathbb{R}$, and the function g represents time-dependent coefficients in an equation generating the process $\{U(t, s)\}$.

There are useful methods of generating a semigroup from a given process other than by (1). (See Dafermos [3] and the references therein.) However, these methods, while having definite advantages over (1) for stability theory, do not improve our results.

3. Some Counterexamples

Perhaps the simplest example of an evolutionary process is when $X = \mathbb{R}$ and each operator $U(t, s)$ is linear and defined on $\mathbb{R} \times \mathbb{R}$. Let $\{U(t, s)\}$ have the form

$$U(t, s)r = e^{g(t, s)}r \tag{2}$$

for some function $g: \mathbb{R} \times \mathbb{R} \to \mathbb{R}$. g satisfies the functional equation

$$g(t + \tau, s) = g(t, \tau + s) + g(\tau, s) \qquad \text{for all} \quad t, \tau, s \in \mathbb{R}. \tag{3}$$

The general solution of (3) is

$$g(t, s) = h(t + s) - h(s), \tag{4}$$

where $h: \mathbb{R} \to \mathbb{R}$ is arbitrary. It is therefore clear that, for example, neither strong measurability nor Baire continuity of $(t, s) \mapsto U(t, s)x$, $x \in X$, suffices to prove continuity of this map when (A) is replaced by an assumption of continuity of $U(t, s)x$ with respect to x alone.

When $\{U(t, s)\} = \{T(t)\}$ is a semigroup then $g(t, s) \equiv f(t)$, where f satisfies Cauchy's equation

$$f(s + t) = f(s) + f(t), \qquad s, t \in \mathbb{R}. \tag{5}$$

By Theorems 1 and 2 any measurable or Baire continuous solution of (5) is continuous, and thus of the form At for some constant A. In 1905, Hamel [5] showed using the axiom of choice that there are discontinuous solutions of (5). Thus even for semigroups of continuous linear operators on a Banach space Theorems 1 and 2 are false without the hypothesis of measurability or Baire continuity on the map $t \mapsto T(t)x$. Can this hypothesis be weakened to the requirement of precompactness of $T((\alpha, \beta))x$ for all $\alpha, \beta \in \mathbb{R}^+$? This question is motivated by the result of Ostrowski [7], who, extending work by Darboux [4] and Sierpinski [8], showed that

JOHN M. BALL

any solution f of (5) that is bounded above on a set of positive measure is necessarily continuous (for an alternative proof, see Kestelman [6]). The answer is no. For example, define $\{T(t)\}$, $t \in \mathbb{R}$, by

$$
\begin{aligned}
T(t)(\pm \pi/2) &= \pm \pi/2 \\
T(t)\tau &= \tan^{-1}[f(t) + \tan \tau], \qquad \tau \in (-\pi/2, \pi/2),
\end{aligned}
\tag{6}
$$

where f is any discontinuous solution of (5). It is easily checked that (6) defines a group of continuous (nonlinear) operators on $[-\pi/2, \pi/2]$ such that each nontrivial orbit is discontinuous [in fact, there will be just one orbit in $(-\pi/2, \pi/2)$ if and only if f is bijective—such solutions f to (5) are easy to construct using a Hamel basis of $\mathbb{R}$ over the rationals]. $\{T(t)\}$, $t \in \mathbb{R}$, can be trivially extended to $\mathbb{R}$. Finally, we remark that $(S(t)\theta)(\tau) = \theta(T(t)\tau)$ for $\theta \in C[0, 1]$ defines a group $\{S(t)\}$, $t \in \mathbb{R}$, of linear isometries on $C[0, 1]$ with the maximum norm such that each nontrivial orbit is discontinuous.

ACKNOWLEDGMENT

This paper was written while the author held part of a United Kingdom Science Research Council research fellowship at the Lefschetz Center for Dynamical Systems, Brown University.

REFERENCES

[1] J. M. Ball, Continuity conditions for nonlinear semigroups, *J. Functional Analysis* **17** (1974), 91–103.
[2] J. M. Ball, Measurability and continuity conditions for nonlinear evolutionary processes, *Proc. Amer. Math. Soc.* (to appear).
[3] C. M. Dafermos, Semiflows associated with compact and uniform processes, *Math. Systems Theory* **8** (1974), 142–149.
[4] M. G. Darboux, Sur le théorème fondamental de la géometrie projective, *Math. Ann.* **17** (1880), 55–61.
[5] G. Hamel, Eine Basis alle Zahlen und die unstetigen Lösungen der Funktionalgleichung: $f(x + y) = f(x) + f(y)$, *Math. Ann.* **60** (1905), 459–462.
[6] H. Kestelman, On the functional equation $f(x + y) = f(x) + f(y)$. *Fund. Math.* **34** (1953), 144–147.
[7] A. Ostrowski, Über die Funktionalgleichung der Exponentialfunktion und verwandte Funktionalgleichungen, *Jahresber. Deut. Math. Verein.* **38** (1929), 54–62.
[8] W. Sierpinski, Sur une propriété des fonctions de M. Hamel, *Fund. Math.* **5** (1924), 334–335.

Stabilization of Linear Evolutionary Processes

RICHARD DATKO

Department of Mathematics
Georgetown University, Washington, D.C.

In this paper we shall indicate the connection between the ability to solve the so-called regulator problem over infinite intervals and the existence of feedback controls that stabilize linear evolutionary processes. Because of space limitations it will be assumed the reader is familiar with the concepts of stability, asymptotic stability, uniform asymptotic stability (see, e.g., [2]), and the basic properties of linear evolutionary processes in a Banach space (see, e.g., [3]). Without further comment we shall consider only processes $S(t, t_0)$ that for $t \geq t_0$ satisfy the conditions

(i) $|S(t, t_0)| \leq M e^{\alpha(t-t_0)}$, where $\alpha > 0$, $M \geq 1$, and both are independent of t_0, and

(ii) $\lim_{t \to t_0^+} S(t, t_0) x_0 = x_0$ for all x_0 in the space.

We shall discuss only one slightly simplified problem, where we consider the interconnection between the solution of the regulator problem and the existence of a stabilizing feedback control.

Problem 1. This involves the optimization over u of the functional

$$C(u, \phi, t_0) = \int_{t_0}^{\infty} (Wx(t, \phi, t_0), x(t, \phi, t_0))\, dt + \int_{t_0}^{\infty} (Uu(t), u(t))\, dt \quad (1)$$

which is subject to the constraint

$$x(t, \phi, t_0) = S(t, t_0)\phi + \int_{t_0}^{t} S(t, s)B(s)u(s)\, ds \quad (2)$$

It is assumed for $t \geq t_0$ that $x(t, \phi, t_0)$ lies in a real Hilbert space H_1, u is a measurable vector-valued function from $[0, \infty)$ into the real Hilbert space H_2, B is a strongly measurable uniformly bounded mapping from $[0, \infty)$ into $\mathcal{L}(H_2, H_1)$, and W and U are, respectively, positive definite mappings in $\mathcal{L}(H_1, H_1)$ and $\mathcal{L}(H_2, H_2)$.

Without further assumptions there is a question concerning the existence of a solution for (1) and (2). This is taken care of if the following hypothesis holds.

Hypothesis 1. For every ϕ in H_1, independently of t_0 there exists a real number $M(\phi)$ such that $C(u, \phi, t_0) \leq M(\phi)$ for at least one measurable $u: [0, \infty) \to H_2$.

If Hypothesis 1 holds then Problem 1 has a unique solution. Furthermore this solution is described by a feedback control that transforms (1) into a uniformly asymptotically stable evolutionary process. To be exact there exists $K: [0, \infty) \to \mathscr{L}(H_1, H_1)$ such that the unique optimal u is given by

$$u(t) = -U^{-1}B^*(t)K(t)x(t) \tag{3}$$

and hence the solution of (1) satisfies a linear integral equation that describes a uniformly asymptotically stable evolutionary process $T(t, t_0)$.

It has been suggested that Hypothesis 1 is a bit strong in that it may be difficult to verify. One answer to this criticism is to indicate the extent to which it is a necessary assumption. For autonomous problems, that is, where B is constant and $S(t, t_0) = S(t - t_0, 0)$ if $t \geq t_0$, the hypothesis is necessary and sufficient if we wish to optimize (1) subject to (2) and ϕ is allowed to run over the entire space H_1. The following is an example of an autonomous system (2) that for $u(t) \equiv 0$ has all solutions tending in norm to zero yet has initial values for which it is impossible to optimize (1). On l_2 consider the control problem $\dot{x} = Ax + Bu$ in which the ith coordinate is described by

$$\dot{x}_i = \begin{cases} -x_i + x_{i+1} + u(t), & 1 \leq i \leq m \\ -x_i + x_{i+1}, & i \geq m+1 \end{cases} \tag{4}$$

Let $\phi = a = (a_1, \ldots, a_n, \ldots)$. Then the solution of (4), $x(t, \phi, u)$, has its ith component given by

$$x_i(t) = \begin{cases} e^{-t}\sum_{n=0}^{\infty} a_{i+n}\dfrac{t^n}{n!} + \sum_{j=1}^{m}\displaystyle\int_0^t e^{-(t-s)}\dfrac{(t-s)^{j-i}}{(j-i)!}u(s)\,ds, & 1 \leq i \leq m \\ e^{-t}\sum_{n=0}^{\infty} a_{i+n}\dfrac{t^n}{n!}, & i \geq m \geq 1 \end{cases} \tag{5}$$

It is shown in [1] that if $u(t) \equiv 0$ the solutions of (4) are asymptotically stable but not uniformly asymptotically stable, and in fact for uniform asymptotic stability, the uncontrolled system must satisfy

$$\int_0^\infty |x(t)|^2 \, dt < \infty$$

for all solutions $x(t)$. However, if $a_i = 0$, $1 \le i \le m$, the solutions of (5) are the same as the uncontrolled system. Hence the system can never satisfy a quadratic functional of the form (1). In fact it can be shown that if $a_i = 1/i$, $i \ge m + 1$, then the value of (1) is infinity for all u. On the other hand if $a_i = 0$, $i \ge m$, then (1) has an optimal solution.

The next two examples indicate that systems that neither are completely controllable nor have homogeneous parts uniformly asymptotically stable may satisfy Hypothesis 1. Consider the mixed initial value–boundary value problem

$$\frac{\partial^2 w}{\partial t^2} = \frac{\partial^2 w}{\partial x^2} - \frac{\partial w}{\partial t} + u(t)$$

$$w(x, 0) = \sum_{n=0}^\infty a_n \cos 2n\pi x = \phi(x) \tag{6}$$

$$\frac{\partial w}{\partial t}(x, 0) = \sum_{n=0}^\infty b_n \cos 2n\pi x = \psi(x)$$

where $\sum_{n=0}^\infty a_n{}^2 + b_n{}^2 < \infty$. The general solution of (6) has the form

$$w(x, t) = \alpha_0 + \beta_0(1 - e^{-t}) + \int_0^t [1 - e^{-(t-s)}]u(s) \, ds$$

$$+ e^{-t/2} \left[\sum_{n=1}^\infty (\alpha_n \cos \omega_n t + \beta_n \sin \omega t) \cos 2n\pi x \right]$$

$$+ e^{-t/2} \sum_{n=1}^\infty \left[\int_0^t (e^{-s/2}/\omega_n) \sin \omega_n(t - s)u(s) \, ds \right] \cos 2n\pi x \tag{7}$$

From (7) we see that if $u(t) \equiv 0$ the solution is stable but not uniformly asymptotically stable. On the other hand, it is easy to see that

$$\alpha_0 + \beta_0(1 - e^{-t}) + \int_0^t [1 - e^{-(t-s)}]u(s) \, ds$$

can be controlled to zero in a finite time T by a control taking only values ± 1. From this it is elementary to verify that Hypothesis 1 holds.

Next consider the second-order system of ordinary differential equations

$$\dot{x}(t) = A(t)x(t) + B(t)u(t) \tag{8}$$

where

$$A(t) = \begin{pmatrix} -1 - \sin t + \cos t & -\sin t + \cos t \\ 1 + \sin t - \cos t & \sin t - \cos t \end{pmatrix}, \qquad B(t) = \begin{pmatrix} \cos t \\ 1 - \cos t \end{pmatrix}$$

and $u(t)$ is a measurable scalar function. If we define for any solution $x(t)$ of (8)

$$y(t) = \begin{pmatrix} \cos t & \cos t \\ 1 - \cos t & 1 - \cos t \end{pmatrix} x(t) \quad \text{and} \quad z(t) = \begin{pmatrix} 1 - \cos t & -\cos t \\ -1 + \cos t & \cos t \end{pmatrix} x(t)$$

then it is elementary to verify $x(t) = y(t) + z(t)$,

$$\dot{y}(t) = A(t)y(t) + B(t)u(t) \qquad \text{and} \qquad \dot{z}(t) = A(t)z(t)$$

Moreover $|z(t)|^2 \leq |z(t_0)|^2 e^{-2(t-t_0)}$, $y(t)$ is completely controllable, and since the system is periodic of period 2π Hypothesis 1 is satisfied.

REFERENCES

[1] Datko, R., Uniform asymptotic stability of evolutionary processes in a Banach space, *SIAM J. Math. Anal.* **3** (1972), 428–445.
[2] Hahn, W., "Stability of Motion." Springer-Verlag, Berlin and New York, 1967.
[3] Krein, S. G., "Linear Differential Equations in a Banach Space." Izdat. Nauk, Moscow, 1967.

Chapter 4: FUNCTIONAL DIFFERENTIAL EQUATIONS

Bifurcation Theory and Periodic Solutions of Some Autonomous Functional Differential Equations

ROGER D. NUSSBAUM *
Department of Mathematics
Rutgers University, New Brunswick, New Jersey

A number of authors [1–8] have studied nonlinear, autonomous functional differential equations and proved the existence of periodic solutions. We shall describe here a new global bifurcation theorem that implies the known existence results for periodic solutions and also provides new information concerning the structure of the set of periodic solutions, e.g., how the period of a periodic solution varies with parameters in the equation. As an illustration, we shall study a van der Pol equation with time lag.

First, we need some definitions. If C is a topological space, $x_0 \in C$, W an open neighborhood of x_0, and $f: W - \{x_0\} \to C$ a continuous map, then x_0 is an "ejective point of f" if there exists an open neighborhood U_0 of x_0 such that for every $x \in U_0 - \{x_0\}$ there is a positive integer $m(x) = m$ such that $f^m(x)$ is defined and $f^m(x) \notin U_0$. If C is a topological space, $y_0 \in C$, W an open neighborhood of y_0, and $f: W \to C$ a continuous map, y_0 is called an "attractive point for f" if there exists an open neighborhood U_0 of y_0 and for any open neighborhood V of y_0 there exists an integer $m(V) = m$ such that $f^j(U_0) \subset W$ for $0 \le j \le m$ and $f^j(U_0) \subset V$ for $j \ge m$.

We now consider the following situation: C is a closed, convex subset of a Banach space X and J is an interval of reals of the form (a, ∞), $-\infty \le a < \infty$. To focus attention on important points, we collect the routine assumptions in one hypothesis.

H1. The origin 0 is an extreme point of C and $0 \notin C$. $F: C \times J \to C$ is a map such that $F(0, \lambda) = 0$ for $\lambda \in J$ and $F|(C - \{0\}) \times J$ is continuous.

---

* Partially supported by NSF GP 43003.

There exists a constant $k < 1$ such that for every bounded set $A \subset C$ and bounded interval $J_0 \subset J$, $F(A \times J_0)$ is bounded and $\gamma(F(A \times J_0)) \leq k\gamma(A)$, where γ denotes measure of noncompactness. There exists a subset Λ of J that has no finite accumulation point and has the property that if J_0 is any compact interval contained in J such that $\Lambda \cap J_0$ is empty, there exists a positive number $\varepsilon = \varepsilon(J_0)$ such that $F(x, \lambda) \neq x$ for $\lambda \in J_0$ and $0 < \|x\| < \varepsilon$.

As usual, it actually suffices in H1 to assume F is a local strict set contraction and $\pi - F$ is a proper map on bounded sets, where $\pi(x, \lambda) = x$. Below we define $F_\lambda(x) = F(x, \lambda)$.

Theorem 1. Suppose that H1 holds. Assume that $\lambda_0 \in \Lambda$ and that F is continuous on an open neighborhood of $(0, \lambda_0)$ in $C \times J$. Assume that there exists an open interval J_0 about λ_0 such that 0 is an attractive point of F_λ for $\lambda \in J_0$, $\lambda < \lambda_0$, and 0 is an ejective point of F_λ for $\lambda \in J_0$, $\lambda > \lambda_0$, or vice versa. Finally, assume that if $a > -\infty$ and if $F(x_j, \lambda_j) = x_j$ for some sequence $(x_j, \lambda_j) \in (C - \{0\}) \times J$ with $\lambda_j \to a$, then $\|x_j\| \to +\infty$. Then if S denotes the closure in $C \times J$ of $\{(x, \lambda) \in C \times J : x \neq 0$ and $F(x, \lambda) = x\}$ and S_0 denotes the maximal closed, connected component of S that contains $(0, \lambda_0)$, it follows that either S_0 is unbounded or $(0, \lambda_1) \in S_0$ for some $\lambda_1 \in \Lambda$ with $\lambda_1 \neq \lambda_0$.

The novelty of Theorem 1 lies in the lack of smoothness assumptions on F_λ at 0 and in the use of some elementary ideas from asymptotic fixed-point theory.

As an application of Theorem 1, consider the following equation:

$$x'(t) = y(t) + \varepsilon x(t) - \alpha(\varepsilon)(x(t))^3, \qquad y'(t) = -x(t - r) - \beta(\varepsilon)(x(t - r))^3 \quad (1)$$

The results we shall describe have analogs for the more general equations

$$x'(t) = y(t) + f_\varepsilon(x(t)), \qquad y'(t) = -g_\varepsilon(x(t - r)) \qquad (2)$$

but for simplicity we shall restrict ourselves to (1). We shall always assume

H2. The constant r is strictly positive and the functions α and β are defined and continuous for all real ε. Furthermore, $\alpha(\varepsilon) > 0$ for all real ε, $\beta(\varepsilon) \geq 0$ for all real ε, and $r\beta(\varepsilon)/\alpha(\varepsilon) < 1$ for all real ε.

As usual, let C denote the cone $\{(\phi, y_0) | y_0 \geq 0; \phi : [-r, 0] \to \mathbb{R}$ is a continuous monotonic increasing function such that $\phi(-r) = 0\}$. For each $(\phi, y_0) \in C$, there exists a unique solution $(x(t), y(t))$ of (1) defined for all $t \geq 0$, and such that $x|[-r, 0] = \phi$ and $y(0) = y_0$. If $(\phi, y_0) \neq 0$, define

$z_1 = z_1(\phi, y_0, \varepsilon)$ to be $\inf\{t > 0 : x(t) = 0\}$. If $z_1 < \infty$, define $z_2 = z_2(\phi, y_0, \varepsilon)$ to be $\inf\{t > z_1 : x(t) = 0\}$. If z_2 is finite, define $F(\phi, y_0, \varepsilon) = F_\varepsilon(\phi, y_0) = (\psi, y_1) \in C$, where $\psi(t) = x(t + z_2 + r)$ for $-r \le t \le 0$ and $y_1 = y(z_2 + r)$. If z_2 is infinite or if $(\phi, y_0) = 0$, define $F(\phi, y_0, \varepsilon) = 0$. It is easy to see that nonzero fixed points (ϕ, y_0) of F_ε give nontrivial periodic solutions of (1) of period $z_2(\phi, y_0, \varepsilon) + r$. One can also prove that $F: C \times \mathbb{R} \to C$ takes bounded sets to precompact sets and is continuous on $[C \times (-\infty, 0)] \cup [(C - \{0\}) \times [0, \infty)]$.

Theorem 2. Assume H2 and suppose that $\lim_{\varepsilon \to -\infty} (\beta(\varepsilon)/\alpha(\varepsilon)) = 0$. Define v_0 to be the unique solution of $v^2 = \cos rv$ such that $0 < v < \pi/2r$ and define $\varepsilon_0 = -(\sin rv_0)/v_0$. Let S denote the closure in $C \times \mathbb{R}$ of $\{((\phi, y_0), \varepsilon) : (\phi, y_0) \ne 0 \text{ and } F_\varepsilon(\phi, y_0) = \phi\}$ and let S_0 denote the maximal connected component of S that contains $(0, \varepsilon_0)$. Then S_0 is unbounded. No point of the form $(0, \varepsilon)$ is an element of S for $\varepsilon \ne \varepsilon_0$. There exists a number $\varepsilon_1 \le \varepsilon_0$ such that $S \cap (C \times (-\infty, \varepsilon_1))$ is empty. Finally, for every finite number ε, $S \cap (C \times (-\infty, \varepsilon))$ is bounded.

Theorem 2 implies immediately that for each $\varepsilon > \varepsilon_0$, Eq. (1) has a nontrivial periodic solution; and even this existence result is new.

For each $((\phi, y_0), \varepsilon) \in S - \{(0, \varepsilon_0)\}$, we can define $p((\phi, y_0), \varepsilon)$ to be the period of the corresponding periodic solution, namely $z_2(\phi, y_0, \varepsilon) + r$; and it is not hard to prove this is a continuous map on $S - \{(0, \varepsilon_0)\}$.

Lemma 1. Assume H2, let v_0 be as in Theorem 2, and define $p(0, \varepsilon_0) = 2\pi/v_0$. Then $p: S \to \mathbb{R}$ is a continuous function on all S.

Lemma 2. Assume H2 and suppose that $\lim_{|\varepsilon| \to \infty} \beta(\varepsilon)/\alpha(\varepsilon) = 0$. Then it follows that

$$\lim_{\varepsilon \to \infty} (\inf\{p(\phi, y_0, \varepsilon') : (\phi, y_0, \varepsilon') \in S \text{ and } \varepsilon' \ge \varepsilon\}) = \infty.$$

Using Lemmas 1 and 2 we immediately obtain

Theorem 3. Assume H2 and suppose that $\lim_{|\varepsilon| \to \infty} \beta(\varepsilon)/\alpha(\varepsilon) = 0$. Let v_0 be as in Theorem 2. Then for each number $q > 2\pi/v_0$, there exists $(\phi, y_0, \varepsilon) \in S_0$ such that $p(\phi, y_0, \varepsilon) = q$. In particular, for each $q > 2\pi/v_0$, there exists an ε and a nonconstant periodic solution of (1) of period precisely q.

The details and proofs of these and related results will appear in [9] and [10].

REFERENCES

[1] R. B. Grafton, A periodicity theorem for autonomous functional differential equations, *J. Differential Equations* **6** (1969), 87–109.

[2] R. B. Grafton, Periodic solutions of certain Lienard equations with delay, *J. Differential Equations* **11** (1972), 519–527.

[3] J. Hale, "Functional Differential Equations." Springer-Verlag, Berlin and New York, 1971.

[4] G. S. Jones, The existence of periodic solutions of $f'(x) = -\alpha f(x-1)[1 + f(x)]$, *J. Math. Anal. Appl.* **5** (1962), 435–450.

[5] G. S. Jones, Periodic motions in Banach space and applications to functional differential equations, *Contrib. Differential Equations* **3** (1964), 75–106.

[6] J. Kaplan and J. Yorke, On the stability of a periodic solution of a differential-delay equation, *SIAM J. Math. Anal.* **6** (1975), 268–282.

[7] R. D. Nussbaum, Periodic solutions of some nonlinear autonomous functional differential equations, *Ann. Mat. Pura Appl.* **101** (1974), 263–306.

[8] R. D. Nussbaum, Periodic solutions of some nonlinear autonomous functional differential equations, II, *J. Differential Equations* **14** (1973), 360–394.

[9] R. D. Nussbaum, A global bifurcation theorem with applications to functional differential equations, *J. Functional Anal.* **19** (1975), 319–339.

[10] R. D. Nussbaum, Global bifurcation of periodic solutions of some autonomous functional differential equations *J. Math. Anal. Appl.* (to appear).

[11] P. Rabinowitz, Some global results for nonlinear eigenvalue problems, *J. Functional Anal.* **7** (1971), 487–513.

A Stability Criterion for Linear Autonomous Functional Differential Equations

F. KAPPEL*

University of Würzburg, Würzburg, Germany

In this paper we report on some results that may be regarded as new details in the general theory of equations cited in the title. To be more precise, we consider the equation

$$\dot{x}(t) = \int_{-h}^{0} d\eta(s)x(t + s), \tag{1}$$

where $\eta(\cdot)$ is an $n \times n$ matrix of bounded variation on $[-h, 0]$, $h > 0$. With respect to notation, definitions, and general results we refer to the book by Hale [2]. If A is the infinitesimal generator of the semigroup defined by the solutions of (1) and λ_0 is in the spectrum

$$\sigma A = P\sigma A = \{\lambda \in \mathbb{C} \mid \det \Delta(\lambda) = 0\}, \qquad \Delta(\lambda) = \lambda I - \int_{-h}^{0} e^{\lambda s}\, d\eta(s),$$

then it is well known that the generalized eigenspace $M_{\lambda_0}(A)$ is given by $\ker(\lambda_0 I - A)^m$, with m the algebraic multiplicity of λ_0, and we have $\dim M_{\lambda_0}(A) = m$ [3]. If we choose a base $\varphi_1, \ldots, \varphi_m$ of $M_{\lambda_0}(A)$ and define $\Phi_{\lambda_0} = (\varphi_1, \ldots, \varphi_m)$ then

$$A\Phi_{\lambda_0} = \Phi_{\lambda_0} B_{\lambda_0},$$

where B_{λ_0} is an $m \times m$ matrix. It is known that the only eigenvalue of B_{λ_0} is λ_0. But it seems that results regarding the Jordan canonical form of B_{λ_0} are not available in the literature. The situation is completely clarified in

Theorem 1. Let λ_0 be in σA with algebraic multiplicity m. Then there exist uniquely determined integers $0 \le d_1 < \cdots < \cdots < d_k$ and $m_1, \ldots, m_k$, $m_j > 0$, such that

$$n = m_1 + \cdots + m_k, \qquad m = d_1 m_1 + \cdots + d_k m_k.$$

* In part supported by the Deutsche Forschungsgemeinschaft, AZ. 477/436/74.

103

Furthermore, the Jordan canonical form of B_{λ_0} contains exactly m_j Jordan cells of dimension d_j, $j = 1, \ldots, k$.

The proof of this theorem is based on the equality

$$\dim(\ker(B_{\lambda_0} - \lambda_0 I)^j) = \dim(\ker \tilde{\Delta}_j), \qquad j = 0, 1, \ldots, \tag{2}$$

where

$$\tilde{\Delta}_j = \begin{pmatrix} \Delta(\lambda_0) & \dfrac{1}{1!}\Delta'(\lambda_0) & \cdots & \dfrac{1}{(j-1)!}\Delta^{(j-1)}(\lambda_0) \\ 0 & & \ddots & \vdots \\ \vdots & \ddots & & \dfrac{1}{1!}\Delta'(\lambda_0) \\ 0 & \cdots & 0 & \Delta(\lambda_0) \end{pmatrix}$$

is a $jn \times jn$ matrix. The rank of $\tilde{\Delta}_j$ can easily be computed if a "local" normal form of $\Delta(\lambda)$ at λ_0 is introduced:

$$K(\lambda) = \mathrm{diag}((\lambda - \lambda_0)^{d_1}K_1(\lambda), \ldots, (\lambda - \lambda_0)^{d_k}K_k(\lambda)), \tag{3}$$

where $K_j(\lambda)$ is $m_j \times m_j$ and $\det K_j(\lambda_0) \neq 0$.

Remark 1. The computation of the numbers d_j and m_j, which are uniquely determined by $\Delta(\lambda)$, can be done according to the following simple algorithm: Define the numbers $\sigma_j = \dim(\ker \tilde{\Delta}_j)$, $j = 1, 2, \ldots$, and $\alpha_0 = n$, $\alpha_1 = \sigma_1$, $\alpha_j = \sigma_j - \sigma_{j-1}$, $j = 2, 3, \ldots$. Then (α_j) is a decreasing sequence and the numbers d_j and m_j are given by

$$\alpha_0 = \cdots = \alpha_{d_1} > \alpha_{d_1+1} = \cdots = \alpha_{d_2} > \alpha_{d_2+1} = \cdots,$$

$$m_j = \alpha_{d_j} - \alpha_{d_j+1}, \qquad j = 1, \ldots, k.$$

Note that $m_k = \alpha_{d_k}$ because $\alpha_{d_k+1} = 0$.

Remark 2. Since the functions φ in $M_{\lambda_0}(A)$ are of the form

$$\varphi(s) = e^{\lambda_0 s} \sum_{j=0}^{d_k} b_{d_k-j} \frac{s^j}{j!}, \qquad s \in [-h, 0],$$

where $y = \mathrm{col}(b_{d_k}, \ldots, b_1)$ is a solution of $\tilde{\Delta}_{d_k} y = 0$, it is not difficult to determine a base

$$\Phi_{\lambda_0}(s) = e^{\lambda_0 s} \sum_{j=0}^{d_k} B_{d_k-j} \frac{s^j}{j!}, \qquad B_i \ \ n \times n \text{ matrices},$$

such that B_{λ_0} is in Jordan canonical form.

Remark 3.   The local normal form $K(\lambda)$ at λ_0 also shows that the order of the pole of $\Delta^{-1}(\lambda)$ at λ_0 is given by $d_k \leq m$. Therefore, d_k is the order of the pole of the resolvent operator $(\lambda I - A)^{-1}$ at λ_0.

Remark 4.   The relations for the numbers d_j, m_j given in Theorem 1 show that $d_k = m$ is possible if and only if either $k = 1$ and $m_1 = n = 1$ or $k = 2$ and $d_1 = 0$, $m_1 = n - 1$, $m_2 = 1$ $(n > 1$, of course). This shows that for a scalar equation

$$x^{(p)}(t) = \sum_{j=0}^{p-1} \int_{-h}^{0} x^{(j)}(t + s) \, d\eta_{j+1}(s),$$

we always have $d_k = m$.

Remark 5.   If we consider the equation adjoint to (1) then λ_0 is in the spectrum of the corresponding infinitesimal generator and the dimension of the corresponding eigenspace is also m. Let $\psi_1, \ldots, \psi_m$ be a base for this eigenspace, ψ_j now being a row vector, and define the $m \times n$ matrix $\Psi_{\lambda_0} = \mathrm{col}(\psi_1, \ldots, \psi_m)$. Then

$$\Psi_{\lambda_0}(s) = e^{-\lambda_0 s} \sum_{j=0}^{d_k-1} C_{d_k-j} \frac{(-s)^j}{j!}, \qquad s \in [0, h],$$

where $(C_1, \ldots, C_{d_k})\tilde{\Delta}_{d_k} = 0$. Ψ_{λ_0} is supposed to be chosen such that $\langle \Psi_{\lambda_0}, \Phi_{\lambda_0} \rangle = I$, where

$$\langle \psi, \varphi \rangle = \psi(0)\varphi(0) - \int_{-n}^{0} \int_{0}^{\theta} \psi(\xi - \theta) \, d\eta(\theta)\varphi(\xi) \, d\xi.$$

The projection φ^{P} of $\varphi \in C([-h, 0], \mathbb{R}^n)$ into $M_{\lambda_0}(A)$ can be calculated using two different ways [1]:

$$\varphi^{\mathrm{P}} = \Phi_{\lambda_0}\langle \Psi_{\lambda_0}, \varphi \rangle,$$

$$\varphi^{\mathrm{P}}(s) = \mathop{\mathrm{Res}}_{\lambda=\lambda_0} (e^{\lambda s}\Delta^{-1}(\lambda)p(\lambda)), \qquad s \in [-h, 0],$$

$$p(\lambda) = \varphi(0) - \int_{-h}^{0} d\eta(s) \int_{0}^{s} e^{\lambda(s-u)}\varphi(u) \, du.$$

If one compares these two expressions for φ^{P}, then the principal part of the expansion of $\Delta^{-1}(\lambda)$ in a Laurent series at λ_0 can be easily obtained if

Φ_{λ_0} is chosen such that B_{λ_0} is in Jordan canonical form, $B_{\lambda_0} = \lambda_0 I + N$:

$$\Delta^{-1}(\lambda) = \frac{1}{(\lambda - \lambda_0)^{d_k}} H_{-d_k} + \cdots + \frac{1}{\lambda - \lambda_0} H_{-1} + \cdots,$$

$$H_{-j} = B_{d_k} N^{j-1} C_{d_k}, \qquad j = 1, \ldots, d_k.$$

Moreover we have

$$\langle \Psi_{\lambda_0}, \Phi_{\lambda_0} \rangle = (C_{d_k}, \ldots, C_1) M \begin{pmatrix} B_{d_.} \\ \vdots \\ B_1 \end{pmatrix}$$

where the $d_k n \times d_k n$ matrix M is given by $M = (M_{ij})$, $i, j = 1, \ldots, d_k$, $M_{ij} = \Delta^{(i+j-1)}(\lambda_0)$.

Remark 6. On the basis of the general theory for linear autonomous functional differential equations the following stability criterion is an obvious consequence of Theorem 1:

Theorem 2. The zero solution of (1) is stable if and only if for all $\lambda_0 \in \sigma A$:

(i) Re $\lambda_0 \leq 0$.
(ii) If Re $\lambda_0 = 0$ and m is the algebraic multiplicity of λ_0, then either

$$m < n \qquad \text{and} \qquad \text{rank } \Delta(\lambda_0) = n - m,$$

or

$$m = n, \qquad \Delta(\lambda_0) = 0, \qquad \text{and} \qquad \text{rank } \Delta'(\lambda_0) = n.$$

If one deals with scalar equations then condition (ii) can be replaced by (see Remark 4)

(ii′) If Re $\lambda_0 = 0$, then λ_0 is a simple root of the characteristic equation.

Remark 7. The results of this chapter also hold for neutral equations. In the case of Theorem 2, the difference operator associated with the equation has to be stable.

REFERENCES

[1] Banks, H. T., and Manitius, A., Projection series for retarded functional differential equations with applications to optimal control problems, *J. Differential Equations* **18** (1975), 296–332.

[2] Hale, J. K., "Functional Differential Equations." Springer, New York, 1971.
[3] Levinger, B. W., A folk theorem in functional differential equations, *J. Differential Equations* **4** (1968), 612–619.

Periodic Differential Difference Equations

JAMES C. LILLO*

Division of Mathematical Sciences
Purdue University, West Lafayette, Indiana

1. The Periodic Case

We consider first the scalar equation

$$\dot{x}(t) = \sum_{j=1}^{n} p_j(t)x(t - \Delta_j),\tag{1.1}$$

where the p_j are continuous real-valued periodic functions of period w and $0 = \Delta_0 < \Delta_1 < \cdots < \Delta_n$. Let $x(t, \zeta, \tau)$ denote the solution of (1.1) that for $t \in [\tau - \Delta_n, \tau]$ has the initial function ζ. Here ζ belongs to $C[-\Delta_n, 0]$, the space of continuous complex-valued functions on $[-\Delta_n, 0]$. In the usual way $x_t = x(t + \theta, \zeta, \tau)$ for $\theta \in [-\Delta_n, 0]$ and $t \geq \tau$ denotes an element of $C[-\Delta_n, 0]$, and the translation map U defined by

$$U x_0 = x_w\tag{1.2}$$

is a compact map of $C[-\Delta_n, 0]$, with the uniform norm, into itself if $w \geq \Delta_n$ [12]. If $w < \Delta_n$, then a suitable power of U is compact and the results are essentially the same [12]. Since U is compact there is for each characteristic multiplier γ an integer $n(\gamma) = \dim$ null space$(U - \gamma I)^m$ for all $m \geq m_0(\gamma)$ and a basis $\phi_1, \ldots, \phi_{n(\gamma)}$ for this null space. For any characteristic multiplier γ and associated basis element ϕ, we shall refer to the solution $x(t, \phi, 0)$ as a solution of Floquet type. We shall assume that the Floquet-type solutions $\{x_i\}$ have been ordered so that if γ_i is the characteristic multiplier associated with x_i and $\lambda_i = (1/w) \log \gamma_i$ is the associated characteristic exponent, then $\operatorname{Re} \lambda_{i+1} \leq \operatorname{Re} \lambda_i$ for all $i \geq 0$. Then it is known [12] that there exists $k_0 > 0$ such that for any $k \geq k_0$ there is an integer $M(k)$ such that for any solution x of (1.1) one has

$$\limsup_{t \to \infty} \left[\left(\ln |x(y)| - \sum_{i=1}^{M(k)} c_i(x)x_i(t) \right) / t \right] < -k.\tag{1.3}$$

* This research was supported by National Science Grant GP 28392.

Equation (1.3) asserts that there is a finite-dimensional subspace in the solution space such that solutions outside this subspace tend to zero at a given exponential rate. The dimension of the subspace depends, of course, on the desired rate of exponential damping. In Section 2 we note how this property can be extended to nonperiodic systems.

The example $\dot{x}(t) = \sin tx(t - 2\pi)$ for which U has only one characteristic multiplier shows that the space of Floquet-type solutions may have finite dimension [12]. For this reason we consider the following two questions:

A.   When does a retarded periodic equation possess a countable set of characteristic exponents?

B.   If there exists a countable set of characteristic exponents, when will the series $\sum_{i=1}^{\infty} c_i(x)x_i(t)$ associated with a given solution x actually converge to $x(t)$?

One may show that it suffices to consider the second question for the Green function $G(t, s)$ [5].

The first results in this direction were obtained by Zverkin [13], who showed that if $\Delta_j = jw$ then the set of characteristic exponents is countable. He also showed that, unless a change of variables of a certain simple form reduced (1.1) to an equation with constant coefficients, then the series $\sum_{i=1}^{\infty} c_i(s)x_i(t)$, associated with $G(t, s)$, diverges for arbitrarily large values of $(t - s)$. In spite of the fact that this divergence is very rapid, it has since been shown [3] that the associated set of eigenfunctions $\{\phi_i\}$ is complete in $C[-\Delta_n, 0]$.

In an attempt to avoid this divergence problem the author first imposed restrictions on the lags Δ_i. Two very limited results were obtained by this means [4, 5]. The next approach was to consider periodic equations as perturbations of autonomous equations. Thus the author [6] considered perturbing the autonomous equation

$$x^{(n)}(t) - c_{0m}x(t - \Delta_m) = 0 \qquad (1.4)$$

by an expression of the form

$$P(x(t)) = \sum_{k=0}^{m-1} \sum_{l=0}^{l(k)} c_{lk}(t)x^{(l)}(t - \Delta_k), \qquad (1.5)$$

where the c_{lk} were all periodic of period w. Since in Eq. (1.4) the translation Δ_m is, roughly speaking, equivalent to n derivatives, one considers the term $x^{(l)}(t - \Delta_k)$ as corresponding to $l + r_k$ derivatives, where $r_k = n\Delta_k/\Delta_m$.

Thus the author established that if

$$\max_k[l(k) + r_k] < n - 2, \tag{1.6}$$

then the periodic equation

$$x^{(n)}(t) - c_{0m}x(t - \Delta_m) = P(x(t)) \tag{1.7}$$

has a countable set of characteristic exponents and the Floquet series associated with the Green function $G(t, s)$ converges for $t - s > 0$.

In later works [7, 8] the author extends this result to more general equations

$$L(x(t)) \equiv x^{(n)}(t) + \sum_{h=0}^{m} \sum_{g=0}^{g(h)} q_{gh} x^{(g)}(t - \sigma_h) = P(x(t)), \tag{1.8}$$

and condition (1.6) on the perturbation term was replaced by a more general condition D, which in the special case of equations of the form (1.7) becomes [8]

$$\max_k[l(k) + r_k] < n - 1. \tag{1.9}$$

The divergence results of Zverkin were then used to show that this result could not be improved except possibly to replace the sign $<$ in (1.9) by $\leq$.

2. The Nonperiodic Case

Condition D mentioned above may be generalized to equations of the form

$$L(x(t)) = P(x(t)), \tag{2.0}$$

where the $c_{lk}(t)$ in $P(x(t))$ are almost periodic or simply continuous functions with uniformly bounded derivatives. The author has shown [9, 10] that these equations possess an approximation property similar to that described in (1.3). Thus for any function one defines the upper characteristic exponents

$$\lambda(f, \pm) = \lim \sup[\log|f(t)|/|t|] \quad \text{as} \quad t \to \pm\infty. \tag{2.1}$$

Then if the c_{lk} are almost periodic, the author has shown [9] that there exists a sequence $\{a_j\}$, $\lim_{j\to\infty} a_j = \infty$, and an increasing sequence $\{W_j\}$ of vector spaces of dimensions $\{M(a_j)\}$. If $y \in W_j$, then y is a solution of (2.0) and

$$\lambda(y, -) \le a_j, \qquad \lambda(y, +) \le a_0. \tag{2.2}$$

Furthermore, every solution x of (2.0) has a unique decomposition

$$x = x_{1j} + x_{2j}, \tag{2.3}$$

where $x_{1j} \in W_j$ and $\lambda(x_{2j}, +) \le -a_j$.

If the c_{lk} are assumed only to be continuous with uniformly bounded derivatives, then the author [10] has shown that for any $\varepsilon > 0$ there is a sequence $\{b_j\}$, $\lim_{j \to \infty} b_j = \infty$, and an increasing sequence $\{W_j\}$ of vector spaces of dimension $\{M(b_j)\}$. If $y \in W_j$, then y is a solution of (2.0) and

$$\lambda(y, -) \le b_j + \varepsilon, \qquad \lambda(y, +) \le b_0 + \varepsilon. \tag{2.4}$$

Furthermore, every solution x of (2.0) may be written in the form

$$x = x_{1j} + x_{2j}, \tag{2.5}$$

where $x_{1j} \in W_j$ and $\lambda(x_{2j}, +) \le -b_j - \varepsilon$.

REFERENCES

[1] V. P. Gurari, N. A. Kulesko, V. I. Macaev, and J. A. Palant, Completeness of the system of Floquet solutions of linear first order differential equations with delays that are multiples of the period of the coefficients, *Vestnik Kharkov Gos. Univ.* **67** (1971), 3–10.

[2] J. Hale, "Functional Differential Equations." Springer, New York, 1971.

[3] N. A. Kuleski, Completeness of the system of Floquet solutions of equations of the neutral type, *Mat. Zemethi* **3** (1968), 297–306.

[4] J. Lillo, Periodic differential difference equations, *J. Math. Anal. Appl.* **15** (1966), 434–441.

[5] J. Lillo, The Green's function for periodic differential difference equations, *J. Differential Equations* **4** (1968), 373–385.

[6] J. Lillo, Periodic differential difference equations of order n, *Amer. J. Math.* **91** (1969), 368–384.

[7] J. Lillo, The Green's function for nth order periodic differential difference equations, *Math. Syst. Theory* **5** (1971), 13–19.

[8] J. Lillo, Periodic perturbations of nth order differential difference equations, *Amer. J. Math.* **94** (1972), 651–676.

[9] J. Lillo, Almost periodic perturbations of nth order differential difference equations, *Math. Syst. Theory* **8** (1974), 207–224.

[10] J. Lillo, Asymptotic behavior of solutions of retarded differential difference equations (to be submitted).

[11] S. Shimanov, The theory of linear differential equations with periodic coefficients and time lag, *P.M.M.* **27** (1963), 450–458.

[12] A. Stokes, A Floquet theory for functional differential equations, *Proc. Nat. Acad. Sci., U.S.A.* **48** (1962), 1330–1334.

[13] A. M. Zverkin, On completeness of a system of characteristic solutions of a differential equation with lags and periodic coefficients, *Trudi Seminar Theory Differential Equations Lags* **2** (1963), Moscow.

Point Data Problems
for Functional Differential Equations

RODNEY D. DRIVER*
Department of Mathematics
University of Rhode Island, Kingston, Rhode Island

Consider, as a simple prototype, the linear scalar delay differential equation

$$x'(t) = f(t)x(t - 1),\tag{1}$$

and let us ask whether it makes any sense to seek solutions on R, or just on $(-\infty, t_0]$, subject to given point data

$$x(t_0) = x_0.\tag{2}$$

At first glance, this problem looks quite hopeless. Indeed, if f is a nonzero constant it is easily seen that there are infinitely many solutions of (1) valid on R that vanish (together with all their derivatives) at t_0 [4].

Doss and Nasr [2] in 1953, and essentially also Fite [5] in 1921, observed that if f is continuous and

$$\int_{-\infty}^{t_0} |f(t)|\, dt < 1,\tag{3}$$

then Eqs. (1) and (2) have a unique *bounded* solution on $(-\infty, t_0]$.

As another example, the functional differential equation

$$x'(t) = f(t)x(g(t)),\tag{4}$$

with nonconstant delay (or advance), has a unique solution satisfying (2) provided f and g are continuous and

$$|g(t) - t_0| \le |t - t_0| \qquad \text{for all} \quad t.\tag{5}$$

This was shown by Fite [5] and was implicit in Polossuchin's dissertation [8] in 1910.

The above authors all considered more general equations than (1) and (4). But their papers are apparently not well known, for minor variants of these

---

* Partially supported by a summer research grant from the University of Rhode Island.

elementary observations have been rediscovered and republished dozens of times in the past few years.

Here we will show how a natural extension of Fite's and Doss and Nasr's theorems does actually have interesting applications.

Let $J \subset R$ be an interval (open, closed, or half open as may be appropriate). Let D be some subset of R^n and let G be a functional mapping $J \times C(J, D) \to R^n$. We write $|\cdot|$ for any convenient norm on R^n. Then we shall consider the functional differential system

$$x'(t) = G(t, x) \tag{6}$$

on J, with given point data

$$x(t_0) = x_0 \tag{7}$$

for some finite $t_0 \in J$ and $x_0 \in D$.

A *solution* of (6) and (7) will mean a differentiable function $x: J \to D$, which satisfies (6) on J and satisfies (7).

Note that this formulation covers systems involving advanced as well as delayed arguments. Note further that the interval J *must be unbounded* whenever constant delays or advances are involved in G.

In order to get uniqueness of the solution of (6) and (7) we will, in general, have to restrict further the class of functions considered (as Doss and Nasr [2], Myškis [6], and others did). Instead of demanding boundedness of x, let us introduce a function

$$\mu \in C(J, (0, 1]),$$

and seek solutions of (6) and (7) in some set

$$S \subset \{z \in C(J, D) : \mu(t)|z(t)| \text{ is bounded}\}.$$

Theorems 1 and 2 give sufficient conditions for existence, uniqueness, and continuous dependence of solutions *in the class S*.

Theorem 1. Assume that

(i)　(S, d) is a complete metric space when

$$d(z, \tilde{z}) \equiv \sup_{t \in J} |z(t) - \tilde{z}(t)|\mu(t).$$

(ii)　For $z \in S$, $G(\cdot, z) \in C(J, R^n)$ and $Tz \in S$, where

$$(Tz)(t) \equiv x_0 + \int_{t_0}^{t} G(s, z)\, ds.$$

(iii)   There exists a function $K_1 \in C(J, R_+)$ such that whenever $z, \tilde{z} \in S$,

$$|G(t, z) - G(t, \tilde{z})| \le K_1(t)d(z, \tilde{z}),$$

while $\mu(t)\left|\int_{t_0}^t K_1(s) \, ds\right| \le r < 1$ for each $t \in J$.

Then (6) and (7) have a unique solution in S.

Proof.   If $z, \tilde{z} \in S$, then $d(Tz, T\tilde{z}) \le rd(z, \tilde{z})$, and the assertion follows from the contraction mapping theorem.

To study continuous dependence of solutions, let G also depend on a parameter λ in some metric space (Λ, ρ). Specifically, let $G : J \times C(J, D) \times \Lambda \to R^n$ and consider the system

$$x'(t) = G(t, x; \lambda) \tag{8}$$

on J with point data (7). Then the following theorem asserts continuous dependence on everything in sight (jointly).

Theorem 2.   Let $N \subset J$ and $H \subset D$, and assume that hypotheses (i) and (ii) of Theorem 1 hold for each $(t_0, x_0, \lambda) \in N \times H \times \Lambda$. Also assume that

(iii')   There exist functions K_1 and $K_2 \in C(J, R_+)$ such that whenever $z, \tilde{z} \in S$, $\lambda, \tilde{\lambda} \in \Lambda$, and $t \in J$,

$$|G(t, z; \lambda) - G(t, \tilde{z}; \tilde{\lambda})| \le K_1(t)d(z, \tilde{z}) + K_2(t)\rho(\lambda, \tilde{\lambda}),$$

while, for each $t_0 \in N$ and $t \in J$,

$$\mu(t)\left|\int_{t_0}^t K_1(s) \, ds\right| \le r < 1, \quad \text{and} \quad \mu(t)\left|\int_{t_0}^t K_2(s) \, ds\right| \le B.$$

Then, given $(t_0, x_0, \lambda) \in N \times H \times \Lambda$ and given $\varepsilon > 0$, there exists $\delta > 0$ such that if $(\tilde{t}_0, \tilde{x}_0, \tilde{\lambda}) \in N \times H \times \Lambda$ with

$$|t_0 - \tilde{t}_0| < \delta, \quad |x_0 - \tilde{x}_0| < \delta, \quad \text{and} \quad \rho(\lambda, \tilde{\lambda}) < \delta,$$

the corresponding unique solutions in S satisfy $d(x, \tilde{x}) < \varepsilon$.

Proof.   Consider (t_0, x_0, λ) a given (fixed) point and $(\tilde{t}_0, \tilde{x}_0, \tilde{\lambda})$ a nearby point, both in $N \times H \times \Lambda$. Then, for each $t \in J$,

$$|x(t) - \tilde{x}(t)| \le |x_0 - \tilde{x}_0| + \left|\int_{t_0}^{\tilde{t}_0} G(s, x; \lambda) \, ds\right|$$

$$+ \left|\int_{t_0}^t K_1(s) \, ds\right| d(x, \tilde{x}) + \left|\int_{t_0}^t K_2(s) \, ds\right| \rho(\lambda, \tilde{\lambda}).$$

Multiply through by $\mu(t)$, recalling that $0 < \mu(t) \le 1$. Then take the supremum over $t \in J$ to find

$$(1 - r)d(x, \tilde{x}) \le |x_0 - \tilde{x}_0| + \left| \int_{t_0}^{t_0} G(s, x; \lambda)\, ds \right| + B\rho(\lambda, \tilde{\lambda}),$$

from which the asserted continuous dependence follows.

The above proofs were very simple because of the nature of the assumptions on G. The theorems provide a format for unifying some special cases, rather than particularly interesting results themselves.

For brevity, only Theorem 1 will be illustrated with examples.

The first two examples, from Nersesjan [7], show the need for the main hypotheses in Theorem 1. Both of these examples use the linear scalar Eq. (1).

Example 1. Consider Eq. (1) with $J = (-\infty, 0]$, $t_0 = 0$, and $f(t) = 2te^{2t-1}$. Apply Theorem 1 with $\mu(t) \equiv 1$, $S = \{z \in C(J, R): z \text{ bounded}\}$, and $K_1(t) = |f(t)|$. It follows that for each $x_0 \in R$ there is one and only one bounded solution of Eqs. (1) and (2) on J. However, it so happens that there is also an unbounded solution given by $x(t) = x_0 e^{t^2}$. See also de Bruijn [1].

Example 2. Consider Eq. (1) with $J = (-\infty, t_0]$, where $t_0 \le 0$, and

$$f(t) = \begin{cases} 0 & \text{for} \qquad t < -1, \\ -2 - 2t & \text{for} \quad -1 \le t \le 0. \end{cases}$$

Again take $\mu(t) \equiv 1$, $S = \{z \in C(J, R): z \text{ bounded}\}$, and $K_1(t) = |f(t)|$. Then for $t_0 < 0$, Theorem 1 guarantees the existence of a unique bounded solution of Eqs. (1) and (2) on J. Moreover, since $f(t) \equiv 0$ for $t < -1$, there can be no unbounded solutions. Thus Eqs. (1) and (2) have a unique solution on J, period!

However, if $t_0 = 0$ condition (iii) of Theorem 1 fails (since $r = 1$). And indeed, it is easily verified that each solution of (1) on $(-\infty, 0]$ has the form

$$x(t) = \begin{cases} c & \text{for} \qquad t < -1, \\ -c(2t + t^2) & \text{for} \quad -1 \le t \le 0, \end{cases}$$

where c is a constant. Thus there are no solutions of (1) and (2) if $x_0 \ne 0$ and infinitely many if $x_0 = 0$.

Example 3 (Small delays). Consider the vector equation

$$x'(t) = F(t, x_t) \qquad \text{on} \quad J = (-\infty, t_0], \tag{9}$$

where F is a continuous functional mapping $J \times \mathcal{C} = J \times C([-\tau, 0], R^n) \to R^n$. We adopt the usual conventions $x_t(s) = x(t + s)$ for $-\tau \le s \le 0$ and $\|\psi\| = \sup_{-\tau \le s \le 0} |\psi(s)|$ for $\psi \in \mathcal{C}$.

Assume there are numbers $A > 0$ and $K > 0$ such that

$$|F(t, 0)| \le Ae^{|t|/\tau} \qquad \text{for} \quad t \le t_0,$$

and whenever (t, ψ) and $(t, \tilde{\psi}) \in J \times \mathcal{C}$,

$$|F(t, \psi) - F(t, \tilde{\psi})| \le K\|\psi - \tilde{\psi}\|.$$

If $K\tau e < 1$, Theorem 1 applies with $\mu(t) = e^{(t-t_0)/\tau}$,

$$S = \{z \in C(J, R^n) : e^{t/\tau}|z(t)| \text{ is bounded}\},$$

and $K_1(t) = Ke^{(t_0 - t)/\tau + 1}$. Thus, for each $x_0 \in R^n$, there is a unique solution of (9) and (7) in S.

Example 3 is of interest because these solutions, extended to R, are actually the "special solutions" that characterize the asymptotic behavior of all solutions of Eq. (9) as $t \to +\infty$. See Rjabov [10], Uvarov [12], and Driver [3].

Corollary (to Example 3). Pointwise degeneracy (see Popov [9]) cannot occur for a system (9) having Lipschitz constant $K < 1/\tau e$ and $F(t, 0)$ bounded.

Example 4 [Generalizing condition (5)]. Consider (6) and (7), with $t_0 = 0$ for convenience. Assume that

(a) $G(\cdot, z) \in C(J, R^n)$ for each $z \in C(J, D)$,

and there exist constants A and K such that

(b) $0 \in D$ and $|G(t, 0)| \le Ae^{K|t|}$ for all $t \in J$,

(c) $|G(t, z) - G(t, \tilde{z})| \le K \sup_{|s| \le |t|, s \in J} |z(s) - \tilde{z}(s)|$, whenever $t \in J$ and $z, \tilde{z} \in C(J, D)$.

Choose any $c > K$ and apply Theorem 1 with $\mu(t) = e^{-c|t|}$,

$$S = \{z \in C(J, D) : e^{-c|t|}|z(t)| \text{ is bounded}\},$$

and $K_1(t) = Ke^{c|t|}$. The conclusion is that Eqs. (6) and (7) have a unique solution in S. Moreover, it can be shown that any solution of (6) and (7) must lie in S. So we actually have unqualified uniqueness.

Example 5.   The equations of one-dimensional motion for two classical electrons give rise to a complicated system of functional differential equations with state-dependent delays [4]. However, in several respects, a reasonable "prototype" appears to be the simple equation

$$y''(t) = a/y^2(t - \tau), \tag{10}$$

where a and τ are positive constants. Let us seek a solution of (10) on $J = (-\infty, 0]$ satisfying

$$y(0) = y_0 > 0, \qquad y'(0) = v_0 < 0. \tag{11}$$

It is convenient to put (10) and (11) into the format of (6) and (7) with $n = 1$. This can be done by introducing $v(t) = y'(t)$ and writing

$$v'(t) = a\left[y_0 - \int_{t-\tau}^{0} v(s)\,ds\right]^{-2} \tag{12}$$

on J, with

$$v(0) = v_0 < 0. \tag{13}$$

By some straightforward estimates, one can show that if (10) and (11) have a solution, then

$$y' = v \in S \equiv \{z \in C(J, R) : v_0 + a/y_0\, v_0 \le z(t) \le v_0\}.$$

Then, using $\mu(t) \equiv 1$, the above set S, and

$$K_1(t) = -2a \cdot (t - \tau)[y_0 + v_0 \cdot (t - \tau)]^{-3},$$

the conditions of Theorem 1 are found to be fulfilled for (12) and (13) *provided* $v_0^2 y_0 > a$. Thus it follows that if $v_0^2 y_0 > a$, then Eqs. (10) and (11) have a unique solution on $(-\infty, 0]$.

With considerably more work, Theorem 1 can be applied to prove the existence of a unique "backwards" solution of the equations of one-dimensional motion for two electrons treated in [4].

Unsolved Problems.   These include Eqs. (10) and (11) on $(-\infty, 0]$ with $y_0 > 0$ and $v_0 = 0$. Existence (but not uniqueness) has recently been proved by Travis [11] for the actual electrodynamics model in [4] for this case (and others).

If this problem has a unique solution, it will then be of interest to study the equations for two charged particles under the influence of both retarded *and advanced* interactions.

Added in Proof: The "unsolved problem" for two electrons described in [4] and mentioned above in connection with Eqs. (10) and (11), with $v_0 = 0$, has now been partially solved. An ingenious proof of existence and uniqueness whenever $v_0^2 + a/y_0$ is sufficiently small and the motion is symmetric was presented by V. I. Zhdanov at the Fourth All-Union Conference on the Theory and Applications of Differential Equations with Deviating Argument, Kiev, USSR, Sept. 1975 (abstracts, pp. 91, 92). In view of the difficulties that Zhdanov had to overcome, Eq. (10) now appears too simple to be a prototype for this problem.

REFERENCES

[1] N. G. de Bruijn, The asymptotically periodic behavior of the solutions of some linear functional equations, *Amer. J. Math.* **71** (1949), 313–330. *MR* **10**-541.

[2] S. Doss and S. K. Nasr, On the functional equation $dy/dx = f(x, y(x), y(x + h))$, $h > 0$, *Amer. J. Math.* **75** (1953), 713–716. *MR* **15**-324.

[3] R. D. Driver, On Ryabov's asymptotic characterization of the solutions of quasi-linear differential equations with small delays, *SIAM Rev.* **10** (1968), 329–341. *MR* **38** #2410.

[4] R. D. Driver, A "backwards" two-body problem of classical relativistic electrodynamics, *Phys. Rev.* **178** (1969), 2051–2057. *MR* **39** #1770.

[5] W. B. Fite, Properties of the solutions of certain functional differential equations, *Trans. Amer. Math. Soc.* **22** (1921), 311–319.

[6] A. D. Myškis, Basic theorems in the theory of ordinary differential equations in nonclassical cases (Russian), *First Math. Summer School, Kanev, 1963: Part* II, pp. 45–116, Izdat. Naukova Dumka, Kiev, 1964. *MR* **32** #7815.

[7] A. B. Nersesjan, A problem for differential–functional equations (Russian), *Akad. Nauk Armjan. SSR Dokl.* **36** (1963), 193–201. *MR* **27** #3856.

[8] O. Polossuchin, Über eine besondere Klasse von differentialen Funktionalgleichungen, Inaugural–Dissertation, Univ. Zürich, 1910.

[9] V. M. Popov, Pointwise degeneracy of linear, time-invariant, delay-differential equations, *J. Differential Equations* **11** (1972), 541–561. *MR* **45** #5515.

[10] Ju. A. Rjabov, Asymptotic properties of the solutions of weakly nonlinear systems with small retardation (Russian), *Trudy Sem. Teor. Differencial. Uravenii s Otklon. Argumentom Univ. Družby Narodov Patrisa Lumumby* **5** (1967), 213–222. *MR* **37** # 545.

[11] S. P. Travis, Existence for a backwards two-body problem of electrodynamics, *Phys. Rev.* D **11** (1975), 292–299.

[12] V. B. Uvarov, Asymptotic properties of solutions of linear differential equations with retarded argument, *Differencial'nye Uravnenija* **4** (1968), 659–663 [*English transl.* pp. 340–342]. *MR* **37** #5506.

Relations between Functional and Ordinary Differential Equations

CARLOS IMAZ

Centro de Investigación y de Estudios Avanzados
del Instituto Politécnico Nacional, México D.F., Mexico

In memoriam: Solomon Lefschetz

In the last few years Zdenek Vorel and the author, and more recently Javier González and Rodolfo Suárez, have been interested in studying the conversion of functional differential equations into ordinary ones. Here we report some of our latest results in this direction. We shall deal only with the so-called retarded case. The advanced and mixed cases are treated in [1].

Let h, a be positive reals. If x is a function from $[-h, a]$ into $\mathbb{R}^n$, the symbol x_t will denote a function, also from $[-h, a]$ into $\mathbb{R}^n$, defined as follows:

$$x_t(s) = \begin{cases} x(s) & \text{if} \quad -h \leq s < t, \\ x(t) & \text{if} \quad t \leq s \leq a. \end{cases}$$

We shall consider the space $L_p([-h, a], \mathbb{R}^n)$, and some subset B of it. We will require of B that if $x \in B$ then x_t makes sense and $x_t \in B$ also. This will be the case if B contains only (classes of) functions that are continuous in $[0, a]$, and then $x(t)$ is the value at t of the continuous element in the class. Of course, we are interested in x_t for $t \in [0, a]$ only.

This being the case, let f be a function

$$f: B \times [0, a] \to \mathbb{R}^n,$$

which is given. Define now a new function F,

$$F: B \times [0, a] \to L_p([-h, a], \mathbb{R}^n)$$

as follows. For any pair $(x, t) \in B \times [0, a]$, $F(x, t)$ is (the class of) the function

$$F(x, t)(\tau) = \begin{cases} 0 & \text{if} \quad -h \leq \tau < t, \\ f(x_t, t) & \text{if} \quad t \leq \tau \leq a, \end{cases}$$

a step function with a jump at t.

Let $\varphi: [-h, a] \to \mathbb{R}^n$ be a given function in B, such that $\varphi(t) = \varphi(0)$ if $0 \le t \le a$.

Consider now the following problems:

$$\dot{x}(t) = f(x_t, t), \qquad 0 \le t \le a,$$
$$x(t) = \varphi(t), \qquad -h \le t < 0, \tag{1}$$

and

$$\dot{y}(t) = F(y(t), t), \qquad 0 \le t \le a,$$
$$y(0) = \varphi. \tag{2}$$

Problem (1) contains the usual retarded functional differential equations, while problem (2) is an ordinary Cauchy problem in L_p.

With these conventions we have:

Theorem 1. If for all $x \in B$ the function $f(x, \cdot) \in L_p([0, a], \mathbb{R}^n)$, then:

(i)   If x is a solution to problem (1) then $y(t) = x_t$ is a solution to problem (2).

(ii)   Conversely, if y is a solution to problem (2), then there exists a unique $x \in B$ such that $y(t) = x_t$, and this x is a solution to problem (1).

The proof of this theorem is based on the following:

Lemma. Under the hypothesis of Theorem 1,

$$\left[\int_0^t F(x, s) \, ds \right](\tau) = \int_0^\beta f(x_s, s) \, ds, \qquad \tau \in [0, a],$$

where $\beta = \min(\tau, t)$.

This lemma is rather nontrivial since it implies that the τ argument can be put inside the left integral. This integral is considered in the sense of Bochner.

The proofs of these results are contained in [2].

As an application we mention the following results, whose proofs are contained in [3].

Theorem 2. If the following hypotheses are fulfilled,

(1) $f(x, \cdot) \in L_p([0, a], \mathbb{R}^n)$,
(2) $\left| \int_0^t [f(x_s, s) - f(y_s, s)] \, ds \right|^p$ is small for $|x - y|_{L_p}$ small, and
(3) $\left| \int_{t_1}^{t_2} f(x_s, s) \, ds \right|^p \le \int_{t_1}^{t_2} m(s) \, ds$, for some integrable m,

then problem (1) has at least one solution.

Theorem 3. If hypotheses (1) and (3) of Theorem 2 are satisfied, and also

(4) $\left| \int_0^t [f(x_s, s) - f(y_s, s)]\, ds \right| \le \int_0^t k(s) |x_s - y_s|_{L_p}$ for some increasing, positive, and integrable k,

then the solution to problem (1) is unique.

The reason these theorems hold is that using the equivalence between problems (1) and (2), as stated in Theorem 1, then Theorem 2 can be reduced to a Carathéodory type of problem, and Theorem 3 is proved by a Gronwall inequality type of argument.

REFERENCES

[1] Suárez, R., Advanced and Ordinary Differential Equations, Doctoral dissertation, Centro de Investigación del IPN, Mexico, 1974.

[2] González, J., Imaz, C., and Vorel, Z., Functional and ordinary differential equations, *Bol. Soc. Mat. Mex.* **18**, 64–69 (1973).

[3] González, J., Doctoral dissertation, Centro de Investigación del IPN, Mexico, 1974.

Asymptotically Autonomous Neutral Functional Differential Equations with Time-Dependent Lag*

A. F. IZÉ

Instituto de Ciências Matemáticas de São Carlos–U.S.P.
Sao Paulo, Brazil

N. A. DE MOLFETTA

Universidade Federal de São Carlos
Sao Paulo, Brazil

The objective of this paper is to extend, for a system of functional differential equations of neutral type, the results obtained by Hale [6] and Cooke [2] for retarded equations and by Izé [8] for neutral equations with constant lag. We give here an idea of the historical development of the problem and the statement of the theorem. The complete proof will appear in the Journal of Mathematical Analysis and Applications. We assume that the reader is familiar with the notation used in the field (see [8] or [5]). Consider

$$\frac{d}{dt}[x(t) - a_0 x(t - r) - a(t)(x(t) - x(t - r))]$$
$$= b_0 x(t) + c_0 x(t - r) + b(t)[x(t) - x(t - r(t))] \tag{1}$$

where $r \geq 0$, a_0, b_0, and c_0 are constants, $a(t)$, $b(t)$ continuous for $t \geq 0$, and $r(t)$ continuous, $0 \leq r(t) \leq r$.

For the retarded differential difference equation

$$\dot{x} = (b_0 + b(t))x(t) + (c_0 + c(t))x(t - r(t)) \tag{2}$$

where $r(t) = r = \text{const}$, Bellman and Cooke [1] showed that if $b(t) \to 0$, $c(t) \to 0$ for $t \to \infty$, and $b(t)$ and $c(t)$ are small in a certain sense, then the solutions of (2) have the form $p(t)e^{\lambda t}$, where $e^{\lambda t}$ is a solution of the unperturbed equation

$$\dot{x} = a_0 x(t) + b_0 x(t - r) \tag{2'}$$

and $p(t)$ is an appropriately selected function of t.

* This research was partially supported by FAPESP, CNPQ, CAPES, and FINEP, Brazil.

Hale [6] considered a more general class of equations of the form

$$\dot{x} = L(x_t) + F(t, x_t) \tag{3}$$

$$x \in C([\sigma - r, \sigma + A], R^n), \qquad x_t(\theta) = x(t + \theta), \qquad F : \Omega \to R^n$$

$$\Omega \in R \times C([-r, 0], R^n), \qquad |F(t, \phi)| \leq \gamma(t)\|\phi\|, \qquad \|\phi\| = \sup_{-r \leq \theta \leq 0} |\phi(\theta)|$$

and was able to extend the results of Bellman and Cooke by proving that if $\int^\infty \gamma(t)\, dt < \infty$ and μ is a simple characteristic root of $\dot{x} = L(x_t)$ such that any other root with real part equal to $\mathrm{Re}\,\mu$ is simple, then Eq. (3) has a solution of the form

$$x(t) = \exp[\mu(t - \sigma) + s(t, \sigma)][a + o(1)],$$

$$s(t, \sigma) = \alpha d \int_0^t F(\tau, \phi_\mu(\tau))\, d\tau, \qquad \phi_\mu(\theta) = e^{\mu\theta}, \qquad -r \leq \theta \leq 0$$

where α and d are given constants.

Cooke [3] considered the equation

$$\dot{x} + ax(t - r)) = 0 \tag{4}$$

or, what is the same, $\dot{x} = -ax(t) + a[x(t) - x(t - r(t))]$, where

$$L(\phi) = -a\phi(0), \qquad F(t, \phi) = a[\phi(0) - \phi(-r(t))]$$

which do not satisfy the hypotheses $\|F(t, \phi)\| \leq \gamma(t)\|\phi\|$, $\int^\infty \gamma(t)\, dt < \infty$, $0 \leq r(t) \leq r$. We can use the space $C([-r, 0], R^n)$, but if ϕ is also Lipschitz, then

$$\|F(t, \phi)\| \leq ak|r(t)| \tag{5}$$

This result shows that we can use a more general hypothesis if we restrict ourselves to a particular subclass of C.

Cooke then considered the subspace C' of C of the continuous Lipschitz functions that satisfy

$$|\phi(\theta_1) - \phi(\theta_2)| \leq k|\theta_1 - \theta_2|, \qquad -r \leq \theta_1 \leq \theta_2 \leq 0 \tag{6}$$

For ϕ in C' we let k_ϕ be the infimum of k for which (6) is valid. We define the norm $\|\phi\|_1 = \max\{k_\phi, \|\phi\|\}$.

Cooke [3] uses the hypothesis $|F(t, \phi)| \leq \gamma(t)\|\phi\|_1$, $\int^\infty \gamma(t)\, dt < \infty$, and extends Hale's paper in [6], but he uses the additional condition $\lim_{t \to \infty} \gamma(t) = 0$ not used by Hale and not used here.

The structure of solutions of Eq. (1) is much more complicated than for a retarded equation. First of all, Eq. (2') has only a finite number of eigenvalues to the right of zero, and if these eigenvalues are simple and have nonpositive real parts, the solutions of (2') are bounded, the operator $T(t)\phi = x_t(\sigma, \phi)$ for (2') is compact, and the spectrum of (2') is enumerable. This is not true for the associated equation

$$\frac{d}{dt}[x(t) - a_0 x(t - r)] = b_0 x(t) + c_0 x(t - r) \tag{1'}$$

The operator $T(t)$ associated to (1') is not compact and also has a continuous spectrum. Additionally, an example given by Gromora and Zverkin [9] shows that, even when (1') has simple characteristic roots with real part less than or equal to zero, (1') may have unbounded solution. Furthermore, the variation of constant formula that represents the solution of (1) is a Stieltjes integral equation.

We extend the result of Hale and Cooke to a general class of neutral equations of the form

$$\frac{d}{dt}[D(x_t) - G(t, x_t)] = L(x_t) + F(t, x_t) \tag{7}$$

$$\frac{d}{dt}[D(x_t)] = L(x_t) \tag{8}$$

where D, L are continuous linear functionals, G is nonatomic at zero [5],

$$D(\phi) = \phi(0) - \int_{-r}^{0} d\mu(\theta)\phi(\theta), \qquad L(\phi) = \int_{-r}^{0} d\eta(\theta)\phi(\theta)$$

with μ, η of bounded variation, $\lim_{\varepsilon \to 0_+} \int_{-\varepsilon}^{0} |d\mu(\theta)| = 0$, and μ has no singular part, that is,

$$\int_{-r}^{0} d\mu(\theta)\phi(\theta) = \sum_{1}^{\infty} A_k \phi(-W_k) + \int_{-r}^{0} A(\theta)\phi(\theta) \, d\theta, \qquad 0 \le W_k \le r$$

$$\sum_{1}^{\infty} |A_k| + \int_{-r}^{0} |A(\theta)| d\theta < \infty, \qquad k = 1, 2, \dots$$

The adjoint equation for (8) is

$$\frac{d}{d\sigma}\left[y(\sigma) - \int_{-r}^{0} y(\sigma - \theta) \, d\mu(\theta)\right] = -\int_{-r}^{0} y(t - \theta) \, d\eta(\theta) \tag{8'}$$

The characteristic values of (8) are the roots of the equation [4, 7]

$$\det \Delta(\lambda) = 0$$

$$\Delta(\lambda) = I - \int_{-r}^{0} e^{\lambda\theta} \, d\mu(\theta) - \int_{-r}^{0} e^{\lambda\theta} \, d\eta(\theta) = \lambda D(e^{\lambda}I) - L(e^{\lambda}I) \tag{9}$$

If $r_D(t) = \exp a_D t$ is the spectral radius of $T(t)$, then $a_D = \sup\{\operatorname{Re} \lambda \,|\, \det \Delta(\lambda)$ $= 0\}$. In particular, if $a_{D_0} = \exp(a_{D_0} t)$ is the spectral radius of $T_{D_0}(t) | C_D$ corresponding to $dD(x_t)/dt = 0$, then $a_{D_0} = \sup\{\operatorname{Re} \lambda \,|\, \det D(e^\lambda I) = 0\}$ if this set is not empty, and $a_{D_0} = -\infty$ otherwise.

If $a > a_{D_0}$, there is only a finite number of roots of $\det \Delta(\lambda) = 0$ with $\operatorname{Re} \lambda \geq a$, and if $\Lambda_a = \{\lambda \in \sigma(A) : \operatorname{Re} \lambda \geq a\}$, then C can be decomposed by Λ_a, as $C = P_a \oplus Q_a$, where P_a and Q_a are subspaces of C invariant under $T(t)$ and A, where

$$A\phi = \lim_{t \to 0_+} \frac{1}{t} [T(t)\phi - \phi]$$

If $x(\sigma, \phi)(t)$ is a solution of (8), then $x_t = x_t^P + x_t^Q$. If $X(t)$ is an $n \times n$ matrix of bounded variation in t for $t \in [0, \infty)$ continuous to the right, for which $D(X_t) = \int_0^t L(X_s)\, ds + I$, then $X_0(\theta) = 0$ if $-r \leq \theta \leq 0$, $X_0(\theta) = 1$ if $\theta = 0$. Since $X(t)$ is a solution of (8), one may take $X_t = T(t)X_0$.

Using the results above and the Banach fixed-point theorem, we prove the following theorem:

Let μ, $\operatorname{Re} \mu > a_{D_0}$, be a simple characteristic root of (8); that is, μ is a simple root of $\det \Delta(\lambda) = 0$. Then there exist a column vector c and a row vector d such that $ce^{\mu t}$ is a solution of (8) and $de^{-\mu \tau}$ is a solution of the adjoint equation (8'), $-\infty < t < \infty$, $-\infty < \tau < \infty$.

Let

$$\phi_\mu(\theta) = ce^{\mu\theta}, \quad -r \leq \theta \leq 0, \qquad \psi_\mu(\tau) = de^{-\mu\tau}, \quad 0 \leq \tau \leq r$$

$$\alpha^{-1} = (\psi_\mu, \Phi_\mu) = \psi_\mu(0)D(\phi_\mu) + \int_{-r}^0 \int_0^\theta \psi_\mu(\xi - \theta)[d\mu(\theta)]\phi_\mu(\xi)\, d\xi \quad (10)$$

$$- \int_{-r}^0 \int_0^\theta \psi_\mu(\xi - \theta)[d\eta(\theta)]\phi_\mu(\xi)\, d\xi$$

and let $X_0(\theta) = 0$, $-r \leq \theta < 0$, $X_0(\theta) = 0$, $\theta = 0$.

Theorem. Suppose that $\operatorname{Re} \mu > a_{D_0}$, where μ is a simple characteristic root of (9), and suppose that all other characteristic roots of (9) with real part equal to $\operatorname{Re} \mu$ are also simple. Let ϕ_μ and α be defined as in (10). Let

$$\delta(t) = \alpha d\{F(t, \phi_\mu) + \mu G(t, \phi_\mu) + F[t, X_0 G(t, \phi_\mu)]\}$$

where $F(t, \phi)$ and $G(t, \phi)$ are linear in ϕ and there exist continuous functions $\gamma(t)$ and $\pi(t)$ for which $F(t, \phi)$ and $G(t, \phi)$ satisfy

$$|F(t, \phi)| \leq \gamma(t)\|\phi\|_1, \qquad |G(t, \phi)| \leq \pi(t)\|\phi\|_1, \qquad t \geq \sigma$$

Let $s(t, \sigma) = \int_0^t \delta(\xi) \, d\xi$ and let $\gamma(t)$ and $\pi(t)$ satisfy one of the following hypotheses:

(I) $\int^\infty \gamma(t) \, dt < \infty$, $\int^\infty \pi(t) \, dt < \infty$, $\lim_{t \to \infty} \pi(t) = 0$;

(II) For any $\beta > 0$ there exists a function $\gamma_2(t)$ defined for $t \geq 0$ for which

$$\lim_{t \to \infty} \pi(t) = 0, \qquad \lim_{t \to \infty} \gamma_2(t) = 0$$

$$\int^\infty \gamma(t)\gamma_2(t) \, dt < \infty, \qquad \int^\infty \pi(t)\gamma_2(t) \, dt < \infty$$

$$\int_\sigma^t \exp[-\beta(t - \tau) - \operatorname{Re} s(t, \tau)]$$

$$\times \, [\gamma(\tau) + \pi(\tau) + \gamma(\tau)\pi(\tau)] \, d\tau \leq \gamma_2(t), \qquad t \geq \sigma \geq 0$$

$$\int_t^\infty \exp[\beta(t - \tau) - \operatorname{Re} s(t, \tau)]$$

$$\times \, [\gamma(\tau) + \pi(\tau) + \gamma(\tau)\pi(\tau)] \, d\tau \leq \gamma_2(t), \qquad t \geq 0$$

$$\int_{t+\theta_2}^{t+\theta_1} \exp[-\beta(t - \tau) - \operatorname{Re} s(t, \tau)][\gamma(\tau) + \pi(\tau) + \gamma(\tau)\pi(\tau)] \, d\tau$$

$$\leq \gamma_2(t)|\theta_1 - \theta_2|, \qquad -r \leq \theta_1, \theta_2 \leq 0$$

Furthermore for any real number v, any n-dimensional row vectors η and λ, and any continuous function W_t defined for $t \geq \sigma$ with values in C' such that $\|W_t\|$ is bounded when $t \to \infty$, we have

$$\left| \int_t^\infty \exp[iv(t - \tau) - s(t, \tau)]\{\eta[F(\tau, W_\tau) + F(\tau, X_0)G(\tau, W_\tau)] - \lambda G(\tau, W_\tau)\} \, d\tau \right|$$

$$\leq [|\eta| + |\lambda|]\gamma_2(t) \sup_{\tau \geq \sigma} \|W_\tau\|_1, \qquad t \geq \sigma$$

Under these conditions, there exists a $\sigma \geq 0$, sufficiently large, and a vector a different from zero such that system (7) has a solution $x(t)$ defined for $t \geq 0$ satisfying

$$x(t) = \exp[\mu(t - \sigma) + s(t, \sigma)][a + o(1)] \qquad \text{when} \quad t \to \infty$$

REFERENCES

[1] Bellman, R., and Cooke, K. L., Asymptotic behavior of solutions of differential-difference equations, *Mem. Amer. Math. Soc.* 35 (1959).

[2] Cooke, K. L., Linear functional differential equations of asymptotically autonomous type, *J. Differential Equations* **7** (1970), 154–174.

[3] Cooke, K. L., Functional differential equations with asymptotically vanishing lag, *Rend. Circ. Mat. Palermo* **16** (1967), 39–56.

[4] Hale, J. K., and Meyer, K., A class of functional differential equations of neutral type, *Mem. Amer. Math. Soc.* **76** (1967), 1–65.

[5] Hale, J. K., Critical cases for neutral functional differential equations, *J. Differential Equations* **10** (1971), 59–82.

[6] Hale, J. K., Linear asymptotically autonomous functional differential equations, *Rend. Circ. Mat. Palermo* **15** (1966), 331–351.

[7] Henry, D., Linear autonomous neutral functional differential equations, *J. Differential Equations* **15** (1974), 106–128.

[8] Izé, A. F., Linear functional differential equations of neutral type asymptotically autonomous, *Ann. Mat. Pura Appl. Ser. IV* **96** (1973), 21–39.

[9] Gromora, P. C., and Zverkin, A. M., On trigonometric series whose sum is a continuous unbounded function on real line and is the solution of an equation with retarded argument (in Russian). *Differentialniya Uraneniya* **4** (1968), 1774–1784.

The Invariance Principle for Functional Equations

N. ROUCHE

Institut de Mathématique
University of Louvain, Louvain-la-Neuve, Belgium

Consider the positive limit set $\Lambda^+(y)$ of a solution $y(t)$ of an autonomous differential equation $y' = f(y)$, whose second member is defined and continuous on some open subset Ω of R^n. A simple and important property is that $\Lambda^+(y) \cap \Omega$ is an invariant set (see, e.g., Hartman [1]). There exist quite a number of generalizations of this property to various types of equations, dynamical systems, and processes (cf. the bibliography of Hale *et al.* [2]). But in most cases, *part* or *all* of the following three hypotheses have been used in the proofs (whereas they do not appear in Hartman's theorem quoted above):

(1)   There is only one solution through every initial point;
(2)   Every solution can be continued up to $-\infty$ on the left and $+\infty$ on the right;
(3)   The solution $y(t)$ generating the limit set either is bounded or remains in some compact, or closed, subset of Ω.

An exception is Strauss and Yorke [3], who need none of these conditions in their study of asymptotically autonomous differential equations.

No doubt at least conditions (2) and (3) are often violated in practical cases. Hence the aim of this paper is to give some answer to the following question: For a retarded differential equation, what "pseudoinvariance" properties of $\Lambda^+(y) \cap \Omega$ can be derived without conditions (1) and (2), and with or without a condition like (3)? Certain types of *positive* pseudo-invariances are easily obtained for unbounded solutions, but it is only by adding the requirement that $y(t)$ be uniformly continuous that we shall be able to ascertain the same types of pseudoinvariances, this time negative as well as positive. And there are indications, to be dealt with elsewhere, that this supplementary hypothesis of uniform continuity cannot be dispensed with completely. Like that of Hartman [1], our proofs, which rely of course on Ascoli's theorem, result from the application of some preliminary theorems on the regularity of solutions. Here we give only the final statements, i.e., those concerning invariance. Detailed proofs, including

133

the regularity theorems, will be published elsewhere. For a study similar to this one, but regarding nonautonomous Carathéodory differential equations, see Rouche [4].

For some $r > 0$, let $C = C([-r, 0], R^n)$ be the space of continuous functions on $[-r, 0]$ into R^n. With no possibility of confusion, we write $\|\cdot\|$ for a norm in R^n and the norm of compact uniform convergence on C. For $b > 0$ and $x: [\sigma - r, \sigma + b] \to R^n$, $t \to x(t)$ a continuous function, we use the standard notation $x_t \in C$, $x_t(\theta) = x(t + \theta)$. Let $D = R \times \Omega$, where Ω is an open set of C. Let $\mathscr{F}$ be the class of continuous functions $f: D \to R^n$. For some $(t_0, \varphi_0) \in D$ and a sequence $(t_0, \varphi_{0k}) \in D$, $k = 1$, 2, ..., we consider the following Cauchy problems:

$$\dot{x} = f(t, x_t), \tag{1}$$

$$x_{t_0} = \varphi_0, \tag{2}$$

$$\dot{x}_k = f_k(t, x_t^k), \tag{3}$$

$$x_{t_0}^k = \varphi_{0k}. \tag{4}$$

We say that the functions f_k, $k = 1, 2, \ldots$, *take closed bounded subsets of D into bounded sets of R^n uniformly with respect to k*, if for every closed bounded subset $F \subset D$, there is an $m > 0$ such that for $k = 1, 2, \ldots$ and every $(t, \varphi) \in F$: $\|f_k(t, \varphi)\| < m$.

The *translate* by a given amount $a > 0$ of a function $f \in \mathscr{F}$ is the function $f_a \in \mathscr{F}$ defined for every $(t, \varphi) \in D$ by $f_a(t, \varphi) = f(t + a, \varphi)$. The following two hypotheses will play a central role hereafter:

(A)   There exists a continuous function $f^*: \Omega \to R^n$ such that for every $(t, \varphi) \in D$, $f_a(t, \psi) \to f^*(\varphi)$ as $a \to \infty$ and $\psi \to \varphi$.

(B)   For every sequence $\{a_k\}$ tending to ∞, there exists a function $f^* \in \mathscr{F}$ and a subsequence $\{a_{k(i)}\}$ such that for every $(t, \varphi) \in D$ and every sequence $\{\varphi_k\}$ tending to φ:

$$f_{a_{k(i)}}(t, \varphi_{k(i)}) \to f^*(t, \varphi) \qquad \text{as} \quad i \to \infty.$$

Remark.   Suppose that φ being fixed, the convergence of $f_{a_{k(i)}}$ to f^*, as mentioned in (B), is for any τ uniform on the half line $[\tau, \infty[$. Then it is known [5] that there exist two functions $g(t, \varphi)$ and $h(t, \varphi)$, the first almost periodic in t in the sense of Bohr, the second approaching zero as $t \to \infty$, and such that $f(t, \varphi) = g(t, \varphi) + h(t, \varphi)$. In this case, f may be called asymptotically almost periodic. Otherwise, of course, the class of functions satisfying property (B) is larger and might well deserve some further exploration.

A subset $F \subset \Omega$ will be said to be *semi-invariant* with respect to Eq. (1), whose second member is supposed to possess property (A) if, for every $(t_0, \varphi_0) \in R \times F$, there exist $\alpha < t_0 - r$, $\omega > t_0$, and a noncontinuable solution $x:]\alpha, \omega[\to R^n$ of the Cauchy problem $\dot{x} = f^*(t, x_t)$, $x_{t_0} = \varphi_0$, such that for every $t \in]\alpha + r, \omega[$, $x_t \in F$.

A subset $F \subset \Omega$ will be said to be *quasi-invariant* with respect to Eq. (1), whose second member is supposed to possess property (B), if for every $(t_0, \varphi_0) \in R \times F$, there exist a limit function f^* of the type mentioned in hypothesis (B), two quantities $\alpha < t_0 - r$, $\omega > t_0$, and a noncontinuable solution $x:]\alpha, \omega[\to R^n$ of the Cauchy problem $\dot{x} = f^*(t, x_t)$, $x_{t_0} = \varphi_0$, such that for every $t \in]\alpha + r, \omega[$, $x_t \in F$.

Positive semi- and *quasi-invariance* are derived from these two concepts, respectively, by substituting for x a solution $x: [t_0 - r, \omega[\to R^n$ noncontinuable to the right.

We are now in a position to state the following theorems.

Theorem 1.   Assume that

(i) f verifies hypothesis (A);

(ii)   For every sequence $\{a_k\} \subset [0, \infty[$, $a_k \to \infty$, the functions f and f_{a_k} take closed bounded sets into bounded sets, uniformly with respect to k; let $x: [t_0 - r, \infty[\to R^n$ be a solution of problems (1) and (2) and assume further that

(iii)   For some closed bounded set $M \subset \Omega$ and all $t \in [t_0, \infty[$, $x_t \in M$, then $\Lambda^+(x)$ is semi-invariant.

Theorem 2.   If one replaces property (A) in Theorem 1 by property (B), then $\Lambda^+(x)$ is quasi-invariant.

Theorem 3.   Assume that

(i) f verifies hypothesis (A);

(ii)   For every sequence $\{a_k\} \subset [0, \infty[$, $a_k \to \infty$, the functions f and f_{a_k} take closed bounded sets into bounded sets uniformly with respect to k; if $x: [t_0 - r, \infty[\to R^n$ is a solution of problems (1) and (2), then $\Lambda^+(x) \cap \Omega$ is positively semi-invariant.

Theorem 4.   If one replaces property (A) in Theorem 3 by property (B), then $\Lambda^+(x) \cap \Omega$ is positively quasi-invariant.

Theorem 5. If one adds to the hypotheses of Theorem 3 that x is uniformly continuous, then $\Lambda^+(x) \cap \Omega$ is semi-invariant.

Theorem 6. If one adds to the hypotheses of Theorem 4 that x is uniformly continuous, then $\Lambda^+(x) \cap \Omega$ is quasi-invariant.

REFERENCES

[1] P. Hartman, "Ordinary Differential Equations." Wiley, New York, 1964.
[2] J. K. Hale, J. P. LaSalle, and M. Slemrod, Theory of a general class of dissipative processes, *J. Math. Anal. Appl.* **39** (1972), 177–191.
[3] A. Strauss and J. A. Yorke, On asymptotically autonomous differential equations, *Math. Syst. Theory* **1** (1967), 175–182.
[4] N. Rouche, The invariance principle applied to noncompact limit sets, *Boll. Un. Mat. Ital.* (to appear).
[5] M. Fréchet, Les fonctions asymptotiquement presque périodiques, *Rev. Sci.* **79** (1941), 341–354.

Existence and Stability of Periodic Solutions of $x'(t) = -f(x(t), x(t-1))$

JAMES L. KAPLAN
Department of Mathematics
Boston University, Boston, Massachusetts

JAMES A. YORKE
Institute for Fluid Dynamics and Applied Mathematics
University of Maryland, College Park, Maryland

During the last twenty-five years there has been a great deal of research on questions of boundedness, stability, and oscillation of solutions to functional differential equations. This research was motivated at first by problems in automatic control, but of late such equations have arisen in the modeling of diverse physical, biological, and ecological phenomena. (See, for example, the recent book of May on ecology [1].) These models often give rise to equations of a very simple type, but which nevertheless exhibit quite complicated behavior. In spite of this, the vast majority of the mathematical literature deals with the local behavior of very general equations. There is a startling lack of results of a nonlinear, global nature for the simplest classes of differential delay equations.

In a recent paper [2], we established a result of this type for the simplest possible differential delay equation,

$$y'(t) = -F(y(t-1)), \tag{1}$$

where $F(0) = 0$, $F: R \to R$ is continuously differentiable, and $dF(y)/dy > 0$ for all $y \in R$. We showed that if $F'(0) > \pi/2$ and $F(y) > -B$ for all $y \in R$, then there is an annulus A in the $(y(t), y(t-1))$ plane whose boundary is a pair of orbits in R^2 of slowly oscillating periodic solutions, and A is asymptotically stable. The region of attraction of A includes all oscillating solutions that do not oscillate too quickly (in the sense that higher harmonics do).

The purpose of this paper is to establish an extension of this result to a broader class of differential delay equations. In particular, we consider equations of the form

$$y'(t) = -f(y(t), y(t-1)), \tag{E}$$

137

where $f(0, 0) = 0, f: R^2 \to R$ is continuously differentiable, and $df(0, y)/dy > 0$. The choice of this particular class of equations was motivated by simple mathematical models that apparently cannot be transformed into the form of Eq. (1). Lasota and Ważewska modeled the size $N(s)$ of the adult red blood cell supply in an animal as a function of time s. The equation that arises in that model is

$$N'(s) = -\mu N(s) + \rho e^{-\gamma N(s-h)}, \qquad s \geq 0, \quad \mu, \rho, \gamma, h > 0. \tag{2}$$

(See [3] for a more complete discussion of the model.) Under a suitable change of variables this equation may be transformed to

$$x'(t) = -ax(t) + e^{-x(t-1)}, \qquad a > 0. \tag{3}$$

Equation (3) has a trivial periodic solution, namely, the constant solution $x(t) \equiv c$, where c is the unique real root of the equation $-ac + e^{-c} = 0$. The change of variables $y = x - c$ now transforms (3) to

$$y'(t) = -ay(t) - ac[1 - e^{-y(t-1)}], \tag{4}$$

for which $y \equiv 0$ is a solution. Numerical studies of Lasota and Ważewska suggested the existence of a nontrivial periodic solution of (4) for small a. These oscillations correspond to blood cell variations observed over periods of weeks in certain trauma cases. Chow [3] was able to use a fixed-point theorem of Browder to establish the existence of such a nontrivial periodic solution, provided a is suitably restricted. We will show that under these hypotheses there exists an asymptotically stable annulus in R^2 whose boundary consists of a pair of nontrivial periodic orbits. All slowly oscillating solutions whose derivative is also slowly oscillating tend asymptotically to this annulus.

Similar results will hold for oscillations around the positive constant solution of the following generalized logistic equation, which occurs in population dynamics,

$$x'(t) = x(t)[a - bx(t) - cx(t - 1)], \qquad a, b, c > 0. \tag{5}$$

(See May [1, pp. 80, 94–100] for cases where either b or c is 0.) Equation (4) is transformable to (E) by letting $y = \ln(x/x_0)$, where $x_0 = a/(b + c)$ is the positive constant solution of (5).

Let us proceed now to a statement of our main result.

Assume that f is continuously differentiable and satisfies the following hypotheses:

(H1) $f(0, 0) = 0,$

(H2) $(\partial f/\partial y)(x, y) > 0$ for all $(x, y) \in R^2,$

(H3) $f(0, y) > -B$ for some $B > 0$,
(H4) $(\partial f/\partial x)(x, y) \geq 0$ for all $(x, y) \in R^2$.

Let $C = C(J, R)$ denote the set of all continuous functions mapping $J = [-1, 0]$ into R. Let C_* denote the set of all $\emptyset \in C$ satisfying

$$\emptyset \text{ has at most one zero on } J, \tag{6}$$

If $\emptyset$ has a zero in $(0, 1)$, then $\emptyset$ must change sign there. $\tag{7}$

If $y(\cdot)$ is a function defined on $[t - 1, t]$ we will write $y_t(\cdot)$ to denote the function $y_t(s) = y(t + s), s \in J$. It was shown in [2] that the set C_* is positively invariant for (1). In that paper, solutions of (1) for which $y_t(\cdot) \in C_*$ for all $t \geq t_0$ were said to be *slowly oscillating*. It is precisely these slowly oscillating solutions that constitute the region of attraction of the asymptotically stable periodic orbits whose existence was established in [2]. Unfortunately, C_* is not necessarily positively invariant for (E), so that we will have to identify a subset C_* for consideration. This subset, which we will denote by $C_*{}^1$, will have the additional property that solutions of (E) for which $y_t(\cdot) \in C_*{}^1$ will have slowly oscillating derivatives. This additional requirement was extraneous in [2] because of the fact that slowly oscillating solutions of (1) automatically possess slowly oscillating derivatives. (See Proposition 2.1 in [2].) Precisely, we have the following.

Definition. Let $C_*{}^1$ denote the set of all those $\emptyset \in C_*$ that satisfy

$$\emptyset(t_1) \neq 0 \quad \text{implies } t_1 \text{ is not a local minimum of } |\emptyset(\cdot)|,$$

$$t_1 \in (-1, 0). \tag{8}$$

Our principal result here will be stated in terms of the behavior of solutions in the $(y(t), y'(t))$ plane. This differs somewhat from our analysis in our previous paper [2], in which we considered solutions in the $(y(t), -y(t - 1))$ plane. The choice of coordinates in that paper was not crucial because of our monotonicity requirements on F in (1). For Eq. (E), however, this new choice of axes results in a considerable simplification of the proofs.

Let $q(t) = (y(t), y'(t))$, where $y(t)$ is a solution of (E). Let $d[\cdot, \cdot]$ denote the usual Euclidean distance.

Definition. Let $S \subset R^2$ be closed. We will say that S is $C_*{}^1$-*stable* in R^2 for (E) if for each $\varepsilon > 0$ there exists a $\delta(\varepsilon) > 0$ such that when y is a slowly oscillating solution of (E) on $[t_0, \infty)$ (i.e., $y_t \in C_*{}^1$ for all $t \geq t_0$) satisfying

$$d[q(t), S] < \delta(\varepsilon) \quad \text{for} \quad t \in [t_0, t_0 + 1],$$

we have

$$d[q(t), S] < \varepsilon \qquad \text{for all} \quad t \geq t_0 .$$

We will say that S is a C_*^1-*global attractor* in R^2 if for each solution y of (E) such that $y_t \in C_*^1$ for some $t \geq 0$ we have $q(t) \to S$ as $t \to \infty$.

We say S is C_*^1-*globally asymptotically stable* if S is C_*^1-stable and is a C_*^1-global attractor.

Definition. For a periodic solution y of (E) we define the *orbit* of y in R^2 by $O_y = \{q(t) : t \in R\}$. We say y is a *simple periodic solution* of (E) if O_y is a simple closed curve.

For a simple periodic solution the curve O_y separates the plane into two parts. We let Int O_y and Ext O_y denote, respectively, the closures of the interior and exterior regions of the plane that are determined by the simple closed curve O_y, in accord with the Jordan curve theorem.

Definition. We say $A \subset R^2$ is a *periodic C_*^1-annulus* if there are simple periodic solutions x and y of (E), with $O_y \subset \text{Int } O_x$ and with $x_t, y_t \in C_*^1$ for all $t \in R$, such that $A = \text{Int } O_x \cap \text{Ext } O_y$.

Definition. We will say that the linear equation

$$y'(t) = -a_1 y(t) - a_2 y(t - 1) \tag{L}$$

is *unstable* if there exists a solution $y(t)$ of (L), defined on $[t_0, \infty)$, for which $\lim_{t \to \infty} \sup |y(t)| = \infty$. For (L), our notion of instability coincides with any of the standard definitions of instability.

We are now prepared to state our principal result.

Theorem 1. Let $f : R^2 \to R$ satisfy (H1)–(H4). Suppose further than (L) is unstable, where $a_1 = (\partial f/\partial x)(0, 0)$ and $a_2 = (\partial f/\partial y)(0, 0)$. Then there exists a periodic C_*^1-annulus $A \subset R^2$ that is C_*^1-globally asymptotically stable for (E).

The techniques we use to prove Theorem 1 are related to the Poincaré–Bendixson method, which we would like to have used in the $(y(t), y'(t))$ plane. Unfortunately, trajectories of solutions of (E) cross in this plane because points in R^2 do not determine solutions uniquely. Our principal lemma (the trajectory crossing lemma) shows that if two trajectories of (E) in R^2 do not cross for a sufficiently large interval of time, then the curves will not cross in the future. This lemma permits the use of analysis similar to that of Poincaré–Bendixson.

REFERENCES

[1] R. M. May, "Stability and Complexity in Model Ecosystems." Princeton Univ. Press, Princeton, New Jersey, 1973.

[2] J. L. Kaplan and J. A. Yorke, On the stability of a periodic solution of a differential delay equation, *SIAM J. Math. Anal.* **6** (1975), 268–282.

[3] S. N. Chow, Existence of periodic solutions of autonomous functional differential equations, *J. Differential Equations* **15** (1974), 350–378.

Existence and Stability of Solutions on the Real Line to $x(t) + \int_{-\infty}^{t} a(t-\tau)g(\tau, x(\tau))\, d\tau = f(t)$, with General Forcing Term

M. J. LEITMAN*
Department of Mathematics and Statistics
Case Western Reserve University, Cleveland, Ohio

V. J. MIZEL**
Department of Mathematics
Carnegie-Mellon University, Pittsburgh, Pennsylvania

We consider the integral equation

$$x(t) + \int_{-\infty}^{t} a(t - \tau)g(\tau, x(\tau))\, d\tau = f(t), \qquad t \in R, \qquad \text{(E)}$$

as well as the related integrodifferential equation

$$\dot{x}(t) + cg(t, x(t)) + \int_{-\infty}^{t} b(t - \tau)g(\tau, x(\tau))\, d\tau = \varphi(t), \qquad t \in R, \qquad \text{(Ė)}$$

for real-valued functions on the line. Here the forcing functions f and φ are restricted only by the requirement that for each $t \in R$, the function f^t (or φ^t) given on the nonnegative axis R^+ by

$$s \mapsto f^t(s) = f(t - s), \qquad s \in R^+,$$

belongs to a preassigned function space F (or Φ) on R^+.

Hereditary equations of type (E) and (Ė) appear in several contexts. From the standpoint of dynamical systems, an equation of the form (E) occurs as a limit equation for the nonlinear Volterra equation

$$y(t) + \int_{0}^{t} a(t - s)h(s, y(s))\, ds = k(t), \qquad t \in R^+, \qquad \text{(V)}$$

---

* Research carried out while Visiting Senior Research Associate at Carnegie-Mellon University, was supported by the National Science Foundation under Grant GP 28377A2.

** Research partially supported by the National Science Foundation under Grant GP 28377A2.

143

that is, functions in the invariant manifold for (V) satisfy (E), for an appropriate pair of functions g, f (see [7–10]). On the other hand, (E) and (Ė) also occur as the description of relations between quantities in theories involving hereditary response and fading memory [1–4, 11]. That is, the functions f and x (respectively, φ and x) are so related that the present value of the output f (respectively, φ) depends on the present value of the input x (respectively, $\dot{x}$) as well as on all past values of x. Such a relationship is often called a hereditary law.

In contrast with (V), the existence problem for (E) does not involve initial data, which are specified by the very form of the equation. Despite the extensive literature regarding (V), there seems to be no previous literature on an existence theory for (E) in the sense that we consider.

Attention is restricted to kernels $a: R^+ \to R$ satisfying

(A_1) a is bounded, nonnegative, and nonincreasing, as well as
(A_2) a is integrable; or
(A_3) for some $\gamma > 0$, $a_\gamma: s \mapsto e^{\gamma s} a(s)$ satisfies (A_1) and (A_2).

Likewise the functions $g: R \times R \to R$ are assumed to satisfy (in addition to Carathéodory conditions):

(G_1) $u g(t, u) \geq 0$, a.e. $t \in R$,

and, sometimes,

(G_2) $u \mapsto g(t, u)$ is isotone a.e. $t \in R$, or
(G_3) for each $M > 0$ there are $\rho_M, T_M > 0$ such that
$$|g(t, u)| \leq \rho_M |u| \quad \text{for} \quad |u| \leq M, \quad \text{a.e.} \quad t \geq T_M.$$

The following main results are established.

(1) If (A_1), (A_2), (G_1) hold then there exists a solution to (E) whenever f^t is continuous and of finite total variation $V^t_{-\infty}(f)$ for each $t \in R$; moreover, at least one solution satisfies
$$|x(t)| \leq |f(t)| + V^t_{-\infty}(f), \qquad t \in R. \tag{$*$}$$

(2) If $b: R^+ \to R$ is bounded, nonnegative, integrable, and of finite first moment, and if (G_1) holds, then there exists a locally absolutely continuous solution to (Ė), with $c = \|b\|_{L^1}$, whenever $\varphi^t: R^+ \to R$ is integrable, for each $t \in R$; moreover, at least one such solution satisfies
$$|x(t)| \leq 2 \int_{-\infty}^{t} |\varphi(\tau)| \, d\tau, \qquad t \in R. \tag{$**$}$$

(3) If (A_1), (A_2), (G_1), (G_2) hold, then there exists a solution to (E) whenever: (a) f' is in $\mathscr{L}^p(R^+)$ $(1 \le p < \infty)$ for each $t \in R$; and (b) on putting $(g \circ f)(t) = g(t, f(t))$, $(g \circ f)'$ is the sum of a function in $\mathscr{L}^1(R^+)$ and a function of finite total variation, for each $t \in R$.

(4) If (A_3), (G_1), (G_2) hold then there exists a solution to (E) whenever f' is in $\mathscr{L}^\infty(R^+)$, for each $t \in R$.

(5) (*Stability*) If (A_1), (A_2), (G_1), (G_3) hold then whenever there exists a solution to (E) for an f that is continuous and of bounded variation on $[T, \infty)$ for some T, x satisfies:

$$0 \le \liminf_{t \to \infty} x(t) \le \limsup_{t \to \infty} x(t) \le f(\infty), \qquad \text{if} \quad f(\infty) \ge 0,$$

$$f(\infty) \le \liminf_{t \to \infty} x(t) \le \limsup_{t \to \infty} x(t) \le 0, \qquad \text{if} \quad f(\infty) < 0.$$

The results obtained above depend on techniques due to Levin [5, 6] together with a body of functional analytic results for hereditary laws developed here and complementing earlier work of Coleman and Mizel [1, 2] and Leitman and Mizel [4]. Although Levin's techniques are restricted to the case of real-valued functions, these functional analytic techniques are not thus restricted.

We close with the statement of two key results on which much of the development depends.

A. Levin's Theorem [5, 6].   Let $a: R^+ \to R$ be bounded, nonnegative, and nonincreasing and let $g: R^+ \times R \to R$ be a Carathéodory function whose Nemytskii operator G takes $\mathscr{L}^\infty_{\text{loc}}(R^+)$ into $\mathscr{L}^1_{\text{loc}}(R^+)$.

(1) If $ug(\tau, u) \ge 0$ a.e. $\tau \in R^+$, then the Volterra equation (V) has a solution for every forcing function f in $\mathscr{C}(R^+) \cap \mathscr{L}\mathscr{B}\mathscr{V}(R^+)$.

(2) If $u \mapsto g(\tau, u)$ is isotone and $\tau \mapsto g(\tau, 0)$ vanishes almost everywhere, then the Volterra equation (V) has a *unique* solution for every forcing function f in $\mathscr{C}(R^+) \cap \mathscr{L}\mathscr{B}\mathscr{V}(R^+)$.

In either case, unique or not, solutions x of (V) are continuous and satisfy the estimate

$$|x(t)| \le |f(t)| + V_0^t(f), \qquad t \in R^+,$$

where $V_0^t(f)$ denotes the total variation of f over the interval $[0, t)$.

B. Theorem. If $a: R^+ \to R$ is nonnegative, nonincreasing, and bounded and $g: R \times R \to R$ preserves measurability under composition, then whenever the measurable function $x: R \to R$ is such that the integral

$$\int_{-\infty}^{t} a(t - \tau)g(\tau, x(\tau)) \, d\tau \triangleq Nx(t)$$

exists for all $t \in R$, Nx is locally absolutely continuous. In addition, the derivative of Nx is given, for almost every $t \in R$, by the formula

$$\dot{\overline{Nx}}(t) = a(0)g(t, x(t)) + \int_{0}^{\infty} g(t - s, x(t - s)) \, da(s).$$

REFERENCES

[1] Coleman, B. D., and Mizel, V. J. Norms and semigroups in the theory of fading memory, *Arch. Rational Mech. Anal.* **23** (1966), 87–123.

[2] Coleman, B. D., and Mizel, V. J., On the general theory of fading memory, *Arch. Rational Mech. Anal.* **29** (1968), 18–31.

[3] Leitman, M. J., and Mizel., V. J., On linear hereditary laws, *Arch. Rational Mech. Anal.* **38** (1970), 46–58.

[4] Leitman, M. J., and Mizel, V. J., On fading memory spaces and hereditary integral equations, *Arch. Rational Mech. Anal.* **55** (1974), 18–51.

[5] Levin, J. J., On a nonlinear Volterra equation, *J. Math. Anal. Appl.* **39** (1972), 458–476.

[6] Levin, J. J., Remarks on a Volterra equation, *in* "Delay and Functional Differential Equations and Their Applications" (K. Schmitt, ed.). pp. 233–255. Academic Press, New York, 1972.

[7] Levin, J. J., and Shea, D. F. On the asymptotic behavior of some integral equations, I, II, III, *J. Math. Anal. Appl.* **37** (1972), 42–82, 288–326, 537–575.

[8] Miller, R. K. "Nonlinear Volterra Integral Equations," Benjamin, New York, 1971.

[9] Miller, R. K., and Sell, G. R., Volterra integral equations and topological dynamics, *Mem. Amer. Math. Soc.* (1970), No. 102.

[10] Sell, G., Nonautonomous differential equations and topological dynamics II. Limiting equations, *Trans. Amer. Math. Soc.* **127** (1967), 263–283.

[11] Weiner, N., "Extrapolation, Interpolation and Smoothing of Stationary Time Series," Wiley, New York, 1949. Reprinted by MIT Press, Cambridge, Massachusetts, 1964.

Existence and Stability for Partial Functional Differential Equations

C. C. TRAVIS

Mathematics Department
The University of Tennessee, Knoxville, Tennessee

G. F. WEBB

Mathematics Department
Vanderbilt University, Nashville, Tennessee

1. Introduction

The purpose of this paper is to investigate existence and stability properties for a class of partial functional differential equations. As a model for this class one may take the equation

$$w_t(x, t) = w_{xx}(x, t) + f(t, w(x, t - r)), \qquad 0 \le x \le \Pi, \quad t \ge 0,$$

$$w(0, t) = w(\Pi, t) = 0, \qquad t \ge 0, \tag{1.1}$$

$$w(x, t) = \phi(x, t), \qquad 0 \le x \le \Pi, \quad -r \le t \le 0,$$

where f is a linear or nonlinear scalar-valued function, r a positive real number, and ϕ a given initial function. In our development the second-derivative term in (1.1) will correspond to a strongly continuous semigroup of linear operators on a Banach space of functions determined by the boundary conditions in (1.1). Accordingly, our approach will rely primarily on semigroup methods and the treatment of (1.1) as an ordinary functional differential equation in a Banach space. Proofs for all results presented may be found in [1].

Before proceeding we shall set forth some notation and terminology that will be used throughout. X will denote a Banach space over a real or complex field; $C = C([-r, 0]; X)$ will denote the Banach space of continuous X-valued functions on $[-r, 0]$ with supremum norm. If μ is a continuous function from $[a - r, b]$ to X and $t \in [a, b]$, then μ_t denotes the element of C given by $\mu_t(\theta) = \mu(t + \theta)$, $-r < \theta < 0$. By a strongly continuous semigroup on X we shall mean a family $T(t)$, $t \ge 0$, of everywhere defined (possibly nonlinear) operators from X to X satisfying $T(t + s) = T(t)T(s)$ for $t, s \ge 0$, and $T(t)x$ is continuous as a function from $[0, \infty)$ to X for each fixed $x \in X$. The infinitesimal generator A_T of $T(t)$ is the function from X to X defined by $A_T x = \lim_{t \to 0^+} (1/t)(T(t)x - x)$ with $D(A_T)$ all x for which this limit exists.

2. Existence of Solutions in the Nonlinear Case

The first existence result is in an integrated form.

Proposition 2.1.   Let $F: [a, b] \times C \to X$ be such that F is continuous and satisfies

$$\|F(t, \psi) - F(t, \hat{\psi})\|_X \le L\|\psi - \hat{\psi}\|_C \qquad \text{for} \quad a \le t \le b, \quad \psi, \hat{\psi} \in C,$$

where L is a positive constant. Let $T(t)$ be a strongly continuous semi-group on X satisfying $|T(t)| \le e^{\omega t}$ for some real number ω. If $\phi \in C$ there is a unique continuous function $\mu(t): [a - r, b] \to X$ that solves

$$\mu(t) = T(t - a)\phi(0) + \int_a^t T(t - s)F(s, \mu_s)\, ds, \qquad a \le t \le b, \quad \mu_a = \phi. \quad (2.1)$$

In the next recent result we place additional hypothesis on F to ensure that $\mu(t)$ satisfies the differential form of (2.1).

Proposition 2.2.   Suppose the hypothesis of Proposition 2.1 and in addition that F is continuously differentiable from $[a, b] \times C$ to X and F_1, F_2 satisfy for $a \le t \le b$, $\psi, \hat{\psi} \in C$, and positive constants β, γ,

$$\|F_1(t, \psi) - F_1(t, \hat{\psi})\|_X \le \beta\|\psi - \hat{\psi}\|_C,$$

$$\|F_2(t, \psi) - F_2(t, \hat{\psi})\|_X \le \gamma\|\psi - \hat{\psi}\|_C.$$

Then for $\phi \in C$ such that $\phi(0) \in D(A_T)$, $\mu(t)$ is continuously differentiable and satisfies

$$(d/dt)\mu(t) = A_T \mu(t) + F(t, \mu_t), \qquad a \le t \le b, \quad \mu_a = \phi. \quad (2.2)$$

3. The Semigroup and Infinitesimal Generator
in the Autonomous Case

Throughout this section we will suppose the hypothesis of Proposition 2.1 except that we require F to be autonomous, that is, $F: C \to X$. By virtue of Proposition 2.1 there exists for each $\phi \in C$ a unique continuous function $\mu(\phi)(t)$: $[-r, \infty) \to X$ satisfying

$$\mu(\phi)(t) = T(t)\phi(0) + \int_0^t T(t-s)F(\mu_s(\phi))\,ds, \qquad t \geq 0, \quad \mu_0(\phi) = \phi. \quad (3.1)$$

For each $t \geq 0$ define $U(t)$: $C \to C$ by $U(t)\phi = \mu_t(\phi)$.

Proposition 3.1. $U(t)$, $t \geq 0$, is a strongly continuous semigroup of (possible nonlinear) operations on C satisfying for $\phi, \hat{\phi} \in C$, $t \geq 0$,

$$\|U(t)\phi - U(t)\hat{\phi}\|_C \leq \|\phi - \hat{\phi}\|_C e^{(\omega+L)t} \qquad\qquad \text{if } \omega \geq 0,$$

$$\|U(t)\phi - U(t)\hat{\phi}\|_C \leq \|\phi - \hat{\phi}\|_C \exp[(\omega + Le^{-\omega r})t] \qquad \text{if } \omega < 0.$$

This semigroup has been extensively studied for ordinary linear functional differential equations by Hale [2] and recently for ordinary nonlinear functional differential equations by Webb [3]. Of particular importance are its compactness properties.

Proposition 3.2. Suppose that $T(t)$ is a compact operator for each $t > 0$. Then $U(t)$ is a compact operator for each fixed $t > r$.

The properties of $U(t)$ permit one to show that the infinitesimal generator of $U(t)$ is defined as follows:

$$(A_U\phi)(\theta) = \dot{\phi}(\theta), \qquad -r \leq \theta \leq 0,$$

$$D(A_U) = \{\phi \in C : \dot{\phi} \in C, \phi(0) \in D(A_T), \dot{\phi}^-(0) = A_T\phi(0) + F(\phi)\}.$$

4. The Spectral Properties of A_U in the Linear Case

Throughout this section F will be as in Section 3 except that we require F to be linear with norm $|F| = L$. $T(t)$ and A_T will be as before except that we require $T(t)$ to be compact for each $t > 0$. $U(t)$ and A_U, which are now linear, will be as in Section 3. For each scalar λ define the linear operator $\Delta(\lambda)$: $D(A_T) \to X$ by

$$\Delta(\lambda)x = A_T x - \lambda x + F(e^{\lambda\theta}x), \qquad x \in D(A_T). \quad (4.1)$$

We will say that λ satisfies the "characteristic equation" of (3.1) provided $\Delta(\lambda)x = 0$ for some $x \neq 0$. The development of the spectral properties of $U(t)$ and A_U follows closely that of Hale [2]. For $t > r$, the spectrum of $U(t)$, $\sigma(U(t))$, is a compact countable set with the only possible accumulation point being zero. Except for the possible point zero, the spectrum of $U(t)$

and the point spectrum of $U(t)$, $P\sigma(U(t))$, coincide and $P\sigma(U(t)) = \exp(tP\sigma(A_U))$. That is, if $\mu = \mu(t) \in P\sigma(U(t))$ for some $t > r$ and $\mu \neq 0$, then there exists $\lambda \in P\sigma(A_U)$ such that $e^{\lambda t} = \mu$. We also have that the point spectrum of A_U coincides with those values of λ that satisfy the characteristic equation (4.1) and that there exists a real number β such that Re $\lambda \leq \beta$ for all $\lambda \in P\sigma(A_U)$. The fundamental result relating the characteristic equation with the growth of solutions of Eq. (3.1) is the following:

Proposition 4.1. Suppose that if λ satisfies the characteristic equation of (3.1), then Re $\lambda \leq \beta$. Then for each $\gamma > 0$ there exists a constant $K(\gamma) \geq 1$ such that for all $t \geq 0$,

$$\|U(t)\phi\|_C \leq K(\gamma)e^{(\beta + \gamma)t}\|\phi\|_C.$$

Let β be the smallest real number such that if λ satisfies the characteristic equation of (3.1), then Re $\lambda \leq \beta$. It follows from the above remarks that if $\beta < 0$, then for all $\phi \in C$, $\|U(t)\phi\|_C \to 0$ as $t \to \infty$. If $\beta = 0$, then there exists $\phi \neq 0 \in C$ such that $\|U(t)\phi\|_C = \|\phi\|_C$ for all $t \geq 0$. If $\beta > 0$, then there exists $\phi \in C$ such that $\|U(t)\phi\|_C \to \infty$ as $t \to \infty$.

5. Stability of Solutions and Examples

Throughout this section we shall consider Eq. (3.1), that is, $U(t)$, $t \geq 0$, will be as in Section 3. We shall call the zero solution $\mu(0)(t)$ of (3.1) stable iff for each $\varepsilon > 0$ there exists $\delta > 0$ such that if $\|\phi\|_C < \delta$, then $\|U(t)\phi\|_C < \varepsilon$ for all $t \geq 0$. We shall call the zero solution $\mu(0)(t)$ of (3.1) asymptotically stable iff it is stable and $\lim_{t \to \infty}\|U(t)\phi\|_C = 0$ for all $\phi \in C$.

Example 5.1. We wish to determine the exact region of stability of the linear equation

$$w_t(x, t) = w_{xx}(x, t) - aw(x, t) - bw(x, t - r), \qquad 0 \leq x \leq \Pi, \quad t \geq 0,$$

$$w(0, t) = w(\Pi, t) = 0, \qquad t \geq 0, \tag{5.1}$$

$$w(x, t) = \phi(t)(x), \qquad 0 \leq x \leq \Pi, \quad -r \leq t \leq 0,$$

as a function of a, b, and r, where the solutions are in the sense of (3.1) for $X = L^2[0, \Pi]$. Let $A_T: X \to X$ be defined by $A_T Y = \ddot{Y}$, $D(A_T) = \{Y \in X : Y \text{ and } \dot{Y} \text{ are absolutely continuous}, \ddot{Y} \in X, Y(0) = Y(\Pi) = 0\}$. Then A_T is the infinitesimal operator of a semigroup $T(t)$, $t \geq 0$, with $\omega = 0$, which is compact for $t > 0$.

The characteristic equation of (5.2) is

$$\Delta(\lambda)f = (A_T - (\lambda + a + be^{-\lambda r})I)f = 0, \qquad f \neq 0 \in D(A_T).$$

Since the eigenvalues of A are $-N^2$, $N = 1, 2, \ldots$, we have from Section 4 that the system is asymptotically stable iff all the roots of the equations

$$\lambda + a + be^{-\lambda r} = -N^2, \qquad N = 1, 2, \ldots,$$

have negative real parts. The exact region of stability of (5.2) as a function of a, b, and r is given in [1].

REFERENCES

[1] C. C. Travis and G. F. Webb, Existence and stability for partial functional differential equations, *Trans. Amer. Math. Soc.* **200** (1974), 395–418.
[2] J. Hale "Functional Differential Equations." Springer-Verlag, Berlin and New York, 1971.
[3] G. F. Webb, Autonomous nonlinear functional differential equations and nonlinear semi-groups, *J. Math. Anal. Appl.* **46** (1974), 1–12.

Periodic Solutions to a Population Equation

JAMES M. GREENBERG
Department of Mathematics
State University of New York at Buffalo, Amherst, New York

In this paper we shall restrict our attention to solutions of the equation

$$x(t) = \gamma \int_{-1}^{-\delta} f(x(t+y))\, dy, \qquad \gamma > 0, \quad 0 \le \delta < 1, \tag{1}_\delta$$

that satisfy $0 < x(t) < 1$. We shall further restrict ourselves to the case where

$$f(x) = x(1-x), \tag{2}$$

although what follows applies equally well to the more general situation:

$$f(0) = f(1) = 0 \quad \text{and} \quad d^2 f/dx^2 < 0, \qquad 0 \le x \le 1. \tag{3}$$

Equation (3) may be used to describe the spread of certain infectious diseases. For details of models leading to $(1)_\delta$ the reader should see [1, 2].

To put our results in perspective, we briefly outline the theory of the equation when $\delta = 0$. These results are due to Cooke and Yorke [2] and Greenberg and Hoppensteadt [3].

The equilibrium points when $\delta = 0$ are $x(t) \equiv 0$ and $x(t) \equiv 1 - 1/\gamma$. The former is asymptotically stable when $0 < \gamma \le 1$, while the latter is when $1 \le \gamma$. Moreover, for $\gamma \gg 1$, the function

$$W(t) \overset{\text{def}}{=} x(t) - 1 + 1/\gamma \tag{4}$$

admits the asymptotic representation

$$W(t) \sim \varepsilon \sum_{n=0} \varepsilon^n W_n(\xi, \tau), \tag{5}$$

where

$$\varepsilon = 1/\gamma, \qquad \xi = (1 - \varepsilon)t, \qquad \tau = \varepsilon^2 t, \tag{6}$$

and W_0 is a well-defined, one-periodic solution of Burger's equation

$$\partial W_0/\partial \tau + [\partial(W_0 - W_0^2)/\partial \xi] - \tfrac{1}{2}\partial^2 W_0/\partial \xi^2 = 0. \tag{7}$$

153

A similar situation persists for $(1)_\delta$ for $0 < \delta < 1$. Here the relevant parameter is

$$\Gamma = \gamma(1 - \delta). \tag{8}$$

The equilibrium points are $x(t) \equiv 0$ and $x(t) \equiv 1 - 1/\Gamma$. The former is asymptotically stable when $0 < \Gamma \le 1$, whereas the latter is when $1 \le \Gamma < \Gamma_c$, $\Gamma_c > 2$. For values of $\Gamma > \Gamma_c$ the constant solution $1 - 1/\Gamma$ ceases to be asymptotically stable; in particular $(1)_\delta$ supports oscillations about $1 - 1/\Gamma$. It is these solutions that we discuss in the remainder of this chapter.

It will be convenient to scale time so that in terms of the new time the minimal period of the oscillation is unity. The scaling is

$$\xi = \omega t, \qquad \omega > 0, \tag{9}$$

and ω must be determined as part of the solution. It will also be convenient to introduce parameters ε and λ defined by

$$\varepsilon = \frac{1 - \delta}{\Gamma - 2} \quad \text{and} \quad \lambda = \delta\omega. \tag{10}$$

The constraints $\Gamma > 2$ and $0 < \delta < 1$ imply that

$$0 < \varepsilon \quad \text{and} \quad 0 < \lambda < \omega. \tag{11}$$

Finally, we let $a > 0$ and $q(\xi, a)$ be defined by

$$aq(\xi, a) = x(\xi/\omega) - 1 + 1/\Gamma, \tag{12}$$

where a is a measure of the amplitude of the difference between $x(\cdot)$ and $1 - 1/\Gamma$, and $q(\cdot, a)$ is normalized so that $\lim_{a \to 0^+} \max_\xi q(\xi, a) = 1$. The problem of finding periodic solutions to $(1)_\delta$ translates into the following problem for $(\lambda, \omega, q(\cdot, a))$: For $a > 0$ and $\varepsilon > 0$ given, find all numbers $0 < \lambda < \omega$ and functions $q(\cdot, a)$ with minimal period equal to unity that satisfy $\lim_{a \to 0^+} \max_\xi q(\xi, a) = 1$ and

$$\varepsilon\omega q(\xi) + \int_{-\omega}^{-\lambda} q(\xi + \eta) \, d\eta + a\left(1 + \frac{2\varepsilon\omega}{\omega - \lambda}\right) \int_{-\omega}^{-\lambda} q^2(\xi + \eta) \, d\eta = 0. \tag{13}$$

In the sequel the word *solution* will refer to a triple $(\lambda, \omega, q(\cdot, a))$ with the above properties.

Our results for this problem are of two types. The first concerns the case $0 < a = \circ(\varepsilon^2)$ and is the content of Theorem 1. The second deals with the case $a = 0(\varepsilon^2)$ and is summarized in Theorem 2.

Theorem 1. Corresponding to each solution of (13) with $a = 0$ there is a solution $(\lambda(a), \omega(a), q(\cdot, a))$ that converges to the given solution of the $a = 0$ equation. These solutions are obtained via the Liapunov–Schmidt method and are well defined for $0 < a \leq \circ(\varepsilon^2)$.

Before we discuss the critical case when $a = 0(\varepsilon^2)$ it is necessary to say a few words about solutions of the $a = 0$ equation. The solutions of $(13)_{a=0}$ with minimal period and maxima equal to one are

$$q(\xi, 0) = \cos 2\pi(\xi - \phi), \qquad \phi \text{ arbitrary}, \tag{14}$$

provided ω satisfies

$$\pi \varepsilon \omega = -\sin 2\pi \omega, \tag{15}$$

$$0 < \lambda \qquad \text{and} \qquad \omega - \lambda = j, \quad \text{a positive integer.} \tag{16}$$

As regards the solutions of (15), we have

Lemma 1. If $\pi N \varepsilon < 1$, then for each $k = 1, 2, \ldots, N$, (15) has no solutions in $[k - 1, \frac{1}{2}(2k - 1)]$ and exactly two solutions ω_{2k-1} and ω_{2k} satisfying

$$\tfrac{1}{2}(2k - 1) < \omega_{2k-1} < \omega_{2k} < k. \tag{17}$$

In what follows we restrict our attention to the roots ω_{2k} and adopt the notation

$$\Omega_k = \omega_{2k}. \tag{18}$$

Lemma 2. The roots Ω_k may be written as

$$\Omega_k = k(1 - D_k), \tag{19}$$

and D_k has the following asymptotic expansion in ε:

$$D_k \sim \frac{\varepsilon}{2} - \frac{\varepsilon^2}{4} + \frac{\varepsilon^3}{48} (6 + (2k)^2 \pi^2) + O(\varepsilon^4). \tag{20}$$

Bearing in mind formulas (16)–(20) we are now in a position to state the main result in the critical case.

Theorem 2. Let $k \geq 1$,

$$a = \tilde{a}_k = \frac{\mu k^2 \varepsilon^2}{24}, \tag{21}$$

$$\omega = \tilde{\Omega}_k = k\left(1 - \frac{\varepsilon}{2} + \frac{\varepsilon^2}{4} - \frac{r_k k^2 \varepsilon^3}{48}\right), \tag{22}$$

$$\lambda = \Lambda_{k,j} = j - \tilde{\Omega}_k, \qquad k \le j \le 2k - 1. \tag{23}$$

Then, for $0 < \varepsilon$ small, (13) has a solution $q_{k,j}(\cdot)$ of the form

$$q_{k,j}(\xi) = q_k(\xi) + \varepsilon v_{k,j}(\xi, \varepsilon), \tag{24}$$

where $q_k(\cdot)$ and r_k satisfy

$$(d^2 q_k/d\xi^2) + (r_k - 6/k^2)q_k + \mu\left(q_k^2 - \int_0^1 q_k^2(s)\, ds\right) = 0,$$

$$q_k(0) = 1 \qquad \text{and} \qquad (dq_k/d\xi)(0) = 0, \tag{26}$$

$$q_k(\cdot) \text{ has minimal period equal to unity.} \tag{27}$$

We shall briefly outline the proof of Theorem 2 when k and hence j are both unity. The equation to be solved is, of course, (13) with $\omega = \tilde{\Omega}_1$, $\lambda = 1 - \tilde{\Omega}_1$, and $a = \tilde{a}_1$. We seek a solution of the form

$$q(\xi) = q_1(\xi) + \varepsilon v_1(\xi, \varepsilon), \tag{28}$$

where q_1 is a C^∞, even, one-periodic function, and $v_1(\cdot, \varepsilon)$ an even, one-periodic function with finite $H_2(0, 1)$ norm. Substitution of (28) into (13) yields the following equation for q_1 and v_1:

$$J_0(q_1) + \varepsilon J_1(q_1, v_1) = 0. \tag{29}$$

The operators J_i are given by

$$J_0(q_1) = \langle q_1, 1 \rangle_0 + \varepsilon^2 c_1(\varepsilon) \|q_1\|_0^2$$
$$- \frac{\varepsilon^3}{24}\left[\frac{d^2 q_1}{d\xi^2} + (r - 6)q_1 + \mu q_1^2\right] + \varepsilon^4 F_4(q_1, r, \mu, \varepsilon), \tag{30}$$

$$J_1(q_1, v_1) = \langle v_1, 1 \rangle_0 + \varepsilon^2 c_1(\varepsilon)\langle 2q_1 v_1 + \varepsilon v^2, 1 \rangle_0$$
$$- \frac{\varepsilon^3}{24}\left[\frac{d^2 v_1}{d\xi^2} + (r - 6 + 2\mu q_1)v_1\right] + \varepsilon^4 G_4(q_1, v_1, r, \mu, \varepsilon), \tag{31}$$

where

$$\langle f, g \rangle_0 = \int_0^1 f(x)g(x)\, dx \qquad \text{and} \qquad \|f\|_0^2 = \langle f, f \rangle_0. \tag{32}$$

The function $q_1(\cdot)$ is now chosen according to formulas (25) and (26).

This problem is solvable and in addition the solution has the property that $\langle q_1, 1 \rangle_0 = 0$. With this choice of q_1, Eq. (29) reduces to

$$\varepsilon(\langle v_1, 1 \rangle_0 + \varepsilon^2 c_1(\varepsilon)\langle 2q_1 v_1 + \varepsilon v_1{}^2, 1 \rangle_0) + \varepsilon^2 \left(c_1(\varepsilon) - \frac{\mu\varepsilon}{24} \right) \|q_1\|_0^2$$

$$\varepsilon^4 \left\{ \frac{1}{24} \left[\frac{d^2 v_1}{d\xi^2} + (r - 6 - 2\mu q_1)v_1 \right] \right.$$

$$\left. - F_4(q_1, r, \mu, \varepsilon) - \varepsilon G_5(q_1, v_1, r, \mu, \varepsilon) \right\} = 0. \tag{33}$$

Equation (33) implies that finding a solution of (13) of the form (28) is equivalent to finding a constant Λ and an even, one-periodic function $v_1(\cdot)$ in $H_2(0, 1)$ such that

$$\frac{d^2 v_1}{d\xi^2} + (r - 6 - 2\mu q_1)v_1 = \Lambda - 24(F_4(q_1, r, \mu, \varepsilon) + \varepsilon G_4(q_1, v_1, r, \mu, \varepsilon)), \tag{34}$$

$$\langle v_1, 1 \rangle_0 + \varepsilon \left(c_1(\varepsilon) - \frac{\mu\varepsilon}{24} \right) \|q_1\|_0^2 + \varepsilon^2 \langle 2q_1 v_1 + \varepsilon v_1{}^2, 1 \rangle_0 - \frac{\varepsilon^3 \Lambda}{24} = 0. \tag{35}$$

That this latter problem has a solution of the desired form for $0 < \varepsilon$ small follows from the contraction mapping principle and from properties of the periodic Green's function for the operator defined by the left-hand side of (34).

REFERENCES

[1] F. Hoppensteadt and P. Waltman, A problem in the theory of epidemics II, *Math. Biosci.* **12** (1971), 133–145.
[2] K. L. Cooke and J. A. Yorke, Some equations modelling growth processes and gonorrhea epidemics, *Math. Biosci.* **16** (1973), 75–101.
[3] J. M. Greenberg and F. Hoppensteadt, Asymptotic behavior of solutions to a population equation, *SIAM J. Appl. Math.* **28** (1975), 662–674.

Existence and Stability of Forced Oscillation in Retarded Equations

ORLANDO LOPES

Campinas, Sao Paulo, Brazil

Notation

(1) $C([a, b], R^n)$ is the Banach space of the continuous functions from the interval $[a, b]$ into R^n, with the norm of the supremum;

(2) If $r \geq 0$ is some fixed number, the space $C([-r, 0], R^n)$ will be denoted simply by C;

(3) If $x \in C([\sigma - r, \sigma + A], R^n)$, $A > 0$, for each $t \in [\sigma, \sigma + A]$ we define $x_t \in C$ putting $x_t(\theta) = x(t + \theta)$, $-r \leq \theta \leq 0$.

If $f: R \times C \to R^n$ is a continuous function, then the relation

$$x(t) = f(t, x_t) \tag{R}$$

is a retarded differential equation. For any $(\sigma, \phi)R \times C$, $x(t, \sigma, \phi)$ denotes the solution of the equation that satisfies $x_\sigma = \phi$. The usual concepts in the stability theory of ordinary differential equations can be extended to retarded differential equations.

Theorem 1. Suppose f is ω-periodic in the t variable and maps bounded sets of $R \times C$ into bounded sets of R^n, and let uniqueness of solution hold. If there is a closed, convex, bounded set K of C that is positively invariant, then there is an ω-periodic solution of Eq. (R) [1].

Example 1. Let us consider the so-called Soper equation for an epidemic [2]:

$$dS(t)/dt = -r(t)S(t)[A\sigma - S(t - \tau) + S(t - \tau - \sigma)] - A, \tag{1}$$

where $S(t)$ is the number of susceptible individuals at time t, A the rate at which new susceptible individuals are recruited to the population, τ the incubation time, and σ the periodic of infection.

Due to seasonal changes, it is reasonable to assume that $r(t)$ is periodic with periodic ω (equal to one year). If we assume that A is positive, it is not difficult to see that if $\phi(\theta) \geq 0$ for $-\tau - \sigma \leq \theta \leq 0$, then $x(t, \phi) \geq 0$

159

for $t \geq 0$, that is, the convex closed set

$$F = \{\phi \in C \ \phi(0) \geq 0, \ -\tau - \sigma \leq \theta \leq 0\}$$

is positively invariant.

Let $S(t)$ be a solution with initial data ϕ satisfying $0 \leq \phi(\theta) \leq R$, $-\tau - \sigma \leq \theta \leq 0$. We are going to analyze under what conditions $S(t) \leq R$ for $t \geq 0$. If t is the first time for which $S(t) = R$, then

$$dS(t)/dt \leq -r_m R(A\sigma - R) + A = r_m R^2 - A\sigma r_m R + A,$$

where $r_m = \min_{0 \leq t \leq \omega} r(t)$. If $r_m > 0$, $A > 0$, and

$$\Delta = A^2 \sigma^2 r_m{}^2 - 4Ar_m = Ar_m(Ar_m \sigma^2 - 4) > 0,$$

that is, for $Ar_m \sigma^2 > 4$, we see that for any R between R_1 and R_2, where R_1 and R_2 are (positive) solutions of $r_m x^2 - A\sigma r_m x + Ar_m x + A = 0$, $dS/dt < 0$, which means that the set $K = \{\phi \in C \ | \ 0 \leq \phi(\theta) \leq R\}$ is positively invariant, and so the equation has an ω-periodic solution. For measles, $\sigma = 2$ days, and the condition above reduces to $Ar_m > 1$.

Definition. Equation (R) is said to be *uniformly ultimately bounded* (U.U.B.) if there are a number R_0 and functions $\alpha(R)$ and $T(R)$ such that for any $\phi \in C$, $|\phi| \leq R$, we have

$$|x_t(t_0, p)| \leq \begin{cases} \alpha(R) & \text{for} \quad t \geq t_0, \\ R_0 & \text{for} \quad t \geq t_0 + T(R), \end{cases}$$

Remark. If uniqueness of solution does not hold, in the definition above $x_t(t_0, \phi)$ means any solution through (t_0, ϕ).

Theorem 2. Assume uniqueness of solution holds for Lipschitzian initial conditions and f maps bounded sets of $R \times C$ into bounded sets of R^n. If Eq. (R) is U.U.B. then it has an ω-periodic solution.

This theorem is a slight modification of a result of Hale [3].

Example 2. We are going to consider the state-dependent lag equation

$$(2) \qquad\qquad \dot{x}(t) = -ax(t - r(x, t)) + f(t),$$

where $0 \leq r(x, t) \leq q$, $r(x, t)$ and $f(t)$ are continuous and ω-periodic in t. We assume also that $r(x, t)$ is locally Lipschitzian in x. This ensures uniqueness of solution for Lipschitzian initial conditions.

Theorem 3. If $0 < aq < \frac{3}{2}$, then Eq. (2) is U.U.B. In particular, it has an ω-periodic solution.

The proof depends on some lemmas.

Lemma 4. If $0 \leq s(t) \leq q$ is a continuous function and $0 < aq < \frac{3}{2}$, then there exist constants K and $\alpha > 0$ that depend just on a and q [not on the function $s(t)$ itself] such that the solution of the *linear* equation

$$\dot{y}(t) = -ay(t - s(t))$$

satisfies

$$|y_t| \leq K \exp[-\alpha(t - t_0)]|y_{t_0}|. \tag{3}$$

Lemma 5. If the solutions of a homogeneous linear equation $\dot{x} = L(t, x_t)$ satisfy

$$|x_t| \leq K \exp[-\alpha(t - t_0)]|x_{t_0}|, \qquad t \geq t_0,$$

then for any bounded continuous function $f(t)$, the solutions of the forced equation

$$\dot{x}(t) = L(t, x_t) + f(t)$$

satisfy

$$|x_t| \leq K \exp[-\alpha(t - t_0)]|x_{t_0}| + \frac{K}{\alpha} \sup_{t_0 \leq u \leq t} |f(u)|, \qquad t \geq t_0. \tag{4}$$

Lemma 4 follows a result of Yorke [4] and Lemma 5 from the variation of constants formula [5].

Proof of Theorem 3. Let $x(t)$ be a solution of (1) and define $s(t) = r(t, x(t))$. Consider the forced *linear* equation

$$\dot{y}(t) = -\alpha y(t - s(t)) + f(t). \tag{L_1}$$

According to Lemmas 4 and 5, we can write

$$|y_t| \leq K \exp[-\alpha(t - t_0)]|y_{t_0}| + \frac{K}{\alpha} \sup_{0 \leq t \leq \omega} |f(t)|, \qquad t \geq t_0,$$

for any solution $y(t)$ of (L_1). However, $x(t)$ *is* a solution of (L_1), and so

$$|x_t| \leq K \exp[-\alpha(t - t_0)]|x_{t_0}| + \frac{K}{\alpha} \sup_{0 \leq t \leq \omega} |f(t)|,$$

which finishes the proof.   Q.E.D.

If R_0 is the number in the definition of U.U.B., let $L = L(R_0)$ be such that

$$| r(x, t) - r(y, t)| \leq L|x - y| \qquad \text{for} \quad |x|, |y| \leq R_0. \tag{5}$$

Using the so-called Razumikhin–Liapunov method we can prove that the following condition is sufficient for global uniform asymptotic stability of the periodic solution:

$$aq < 1/1 + L(\alpha R_0 + F), \tag{6}$$

where $F = \sup |f(u)|$, $u \in R$.

The following estimate for R_0 can be obtained:

$$R_0 < F(1 + aq)/(1 - aq), \qquad \text{if} \quad aq < 1.$$

Using a result of Miller [6] we can prove that if condition (6) holds and $r(x, y)$ and $f(t)$ are almost periodic in t, then there is an almost-periodic solution.

REFERENCES

[1] Hale, J., and Lopes, O., Dissipative processes and asymptotic fixed point theorems, *J. Differential Equations* **13** (1973), 391–402.

[2] Wilkins, J. E., The differential–difference equation for epidemics, *Bull. Math. Biophys.* **7** (1945), 149–150.

[3] Hale, J., An asymptotic fixed point theorem for retarded equations, unpublished.

[4] Yorke, J. A., Asymptotic stability for one dimensional differential–delay equation, *J. Differential Equations* **1** (1970), 189–202.

[5] Hale, J., "Functional Differential Equations" (*Appl. Math. Sci.*), Vol. 3. Springer-Verlag, Berlin and New York, 1971.

[6] Miller, R. K., Almost periodic differential equations, *J. Differential Equations* **11** (1965), 337–345.

Exact Solutions of Some Functional Differential Equations

CLEMENT A. W. McCALLA
Department of Mathematics
Howard University, Washington, D.C.

1. Representations of Solutions of Linear Functional Differential Equations

Representations of solutions of linear functional differential equations can be found in Banks [1] and Delfour and Mitter [2]. In particular, from [2] we have that the solution to the scalar equation

$$\dot{x}(t) = \sum_{i=0}^{N} A_i x(t + \theta_i) + \int_{-a}^{0} A(\theta)x(t + \theta)\, d\theta + f(t), \qquad t > 0,$$

$$x(t) = h(t), \qquad -a \le t \le 0,$$

(1)

where

$$-a = \theta_N < \cdots < \theta_1 < \theta_0 = 0, \qquad A_i = \text{const}, \quad i = 0, \ldots, N,$$

$$A \in L^1(-a, 0), \qquad f \in L^1_{\text{loc}}(0, \infty), \qquad h \in L^p(-a, 0), \quad p \ge 1,$$

where $h(0)$ is given, is provided by

$$x(t) = \Phi(t)h(0) + \sum_{i=1}^{N} \int_{\theta_i}^{\min(0,\, t+\theta_i)} d\alpha\, \Phi(t - \alpha + \theta_i)A_i h(\alpha)$$

$$+ \int_{-a}^{0} d\alpha \int_{\max(-a,\, \alpha - t)}^{\alpha} d\beta\, \Phi(t - \alpha + \beta)A(\beta)h(\alpha) + \int_{0}^{t} \Phi(t - s)f(s)\, ds, \quad (2)$$

where $\Phi(t)$, the Green's function, satisfies (1) with $f = 0$ and the initial condition

$$\Phi(t) = 0, \quad t < 0, \qquad \Phi(0) = 1.$$

In the case where $A_i = 0$, $i = 0, \ldots, N$, define

$$\mathscr{A}(t) = \begin{cases} 0, & t < 0, \\ A(-t), & 0 \le t \le a, \\ 0, & t > 0. \end{cases}$$

163

Then $\Phi(t)$ satisfies the Volterra integrodifferential equation of convolution type

$$\dot{x}(t) = \int_0^t \mathscr{A}(s)x(t-s)\,ds, \quad t > 0, \qquad x(0) = 1. \tag{3}$$

In other words, *the Green's function of a linear autonomous retarded functional differential equation satisfies an integrodifferential equation of convolution type with kernel of compact support.*

If $A_i \neq 0$ for some i, then we construct a function of bounded variation $\mathscr{A}(t)$ as follows: $\mathscr{A}(t) = 0$ if $t \leq 0$; $\mathscr{A}(t)$ is absolutely continuous on $(-\theta_i, -\theta_{i+1})$ with derivative $A(-t)$ and with jumps A_i at $t = -\theta_i$; $\mathscr{A}(t)$ is constant for $t \geq a$. Then $\Phi(t)$ satisfies the Volterra–Stieltjes integrodifferential equation of convolution type

$$\dot{x}(t) = \int_0^t d\mathscr{A}(s)x(t-s), \quad t > 0, \qquad x(0) = 1. \tag{4}$$

2. Solution of Integrodifferential Equations

Consider

$$\dot{x}(t) = \int_0^t \mathscr{A}(s)x(t-s)\,ds, \quad t > 0, \qquad x(0) = 1, \tag{5}$$

where

$$\mathscr{A} \in L^1_{\text{loc}}(0, \infty).$$

Define a sequence $\mathscr{A}_n$ by the relation $\mathscr{A}_n = \mathscr{A}_1 * \mathscr{A}_{n-1}$ for $n > 1$ and define $\mathscr{A}_1$ to be $\mathscr{A}$. Here $f * g$ denotes the convolution of the two functions f and g.

Theorem 1. The solution of (5) is given by

$$\Phi(t) = 1 + \sum_{n=1}^{\infty} \frac{1}{n!} \int_0^t \mathscr{A}_n(s)(t-s)^n\,ds. \tag{6}$$

Proof. Let $T < \infty$ and $\int_0^T |\mathscr{A}_1(s)|\,ds = c$. Then $\int_0^T |\mathscr{A}_n(s)|\,ds \leq c^n$, and $|\Phi(t)| \leq e^{cT}$ for $t \in [0, T]$. Hence, the series (6) converges absolutely and uniformly for $0 \leq t \leq T$. The series (6) can be differentiated term by term since the series of derivatives

$$\Phi'(t) = \sum_{n=1} \frac{1}{(n-1)!} \int_0^t \mathscr{A}_n(s)(t-s)^{n-1} \, ds,$$

$$|\Phi'(t)| \leq c e^{cT}, \qquad \text{for} \quad 0 \leq t \leq T, \tag{7}$$

is uniformly convergent on $0 \leq t \leq T$.

The series (6) satisfies the initial condition of (5) and differentiating term by term it can be shown to satisfy the differential equation (5). Hence, (6) is a solution of (5). The uniqueness of solutions of (5) can be established by applying Gronwall's inequality to an integral version of (5). Hence (6) is the solution to (5).

Consider

$$\dot{x}(t) = \int_0^t d\mathscr{A}(s)x(t-s), \quad t > 0, \qquad x(0) = 1, \tag{8}$$

where

$$\mathscr{A} \in BV_{\text{loc}}(0, \infty).$$

Define a sequence of measures $d\mathscr{A}_n$ by the relation $d\mathscr{A}_n = d\mathscr{A}_1 * d\mathscr{A}_{n-1}$ for $n > 1$ and define $d\mathscr{A}_1$ to be $d\mathscr{A}$. Here $d\mu * dv$ denotes the convolution of the two measures $d\mu$ and dv.

Theorem 2. The solution of (8) is given by

$$\Phi(t) = 1 + \sum_{n=1} \frac{1}{n!} \int_0^t d\mathscr{A}_n(s)(t-s)^n. \tag{9}$$

Remark. On the practical side, it may be preferable to compute the convolution of functions and distributions rather than of measures. In computing the Green's function $\Phi(t)$ for Eq. (1), if we let

$$\mathscr{A}_1(t) = A(-t) + \sum_{i=1}^N A_i \, \delta(t + \theta_i)$$

be the zero-order distribution, then $\mathscr{A}_n(t)$ is also a zero-order distribution corresponding to the nth convolution of the measure, and we can use expression (6) rather than expression (9).

3. Examples

Example 1

$$\dot{x}(t) = Ax(t), \qquad \mathcal{A}(t) = A\,\delta(t), \qquad \Phi(t) = 1 + \sum_{n=1}^{\infty} \frac{1}{n!}\,A^n t^n = \exp At.$$

Example 2

$$\dot{x}(t) = \sum_{i=0}^{N} A_i\, x(t + \theta_i), \qquad \mathcal{A}(t) = \sum_{i=0}^{N} A_i\,\delta(t + \theta_i),$$

$$\Phi(t) = 1 + \sum_{n=1}^{\infty} \sum_{r_0 + \cdots + r_N = n} \frac{1}{r_0! \cdots r_N!}$$
$$\times A_0^{r_0} \cdots A_N^{r_N}(t + r_0\theta_0 + \cdots + r_N\theta_N)^n H(t + r_0\theta_0 + \cdots + r_N\theta_N),$$

where $H(t)$ is the unit Heaviside step function and $r_i \geq 0$, $i = 0, \ldots, N$. Note that for fixed t, we have a finite sum if $A_0 = 0$.

Example 3

$$\dot{x}(t) = \int_{-a}^{0} A(\theta)x(t + \theta)\,d\theta, \qquad A(\theta) = \alpha_j, \quad \theta \in (\theta_j, \theta_{j-1}), \quad j = 1, \ldots, N.$$

Let

$$A_0 = \alpha_1, \qquad A_j = (\alpha_{j+1} - \alpha_j), \qquad j = 1 \ldots N - 1, \quad A_N = -\alpha_N.$$

Then

$$\Phi(t) = 1 + \sum_{n=1}^{\infty} \sum_{r_0 + \cdots + r_N = n} \frac{n!}{r_0! \cdots r_N!} \frac{1}{2n!}$$
$$\times A_0^{r_0} \cdots A_N^{r_N}(t + r_0\theta_0 + \cdots + r_N\theta_N)^{2n} H(t + r_0\theta_0 + \cdots + r_N\theta_N).$$

Again, note finite summation for the case $A_0 = 0$.

Example 4

$$\dot{x}(t) = \int_{0}^{t} Ks^m x(t - s)\,ds, \qquad m \text{ positive integer},$$

$$\Phi(t) = 1 + \sum_{n=1}^{\infty} \frac{K^n (m!)^n}{(mn + 2n)!}\, t^{mn + 2n}.$$

If $m = 0$,

$$\Phi(t) = \begin{cases} \cosh\sqrt{K}\,t, & K > 0, \\ \cos\sqrt{-K}\,t, & K < 0. \end{cases}$$

4. Generalizations

The Green's function corresponding to a nonautonomous functional differential equation will satisfy an integrodifferential equation not of convolution type.

Let us consider the Volterra integrodifferential equation

$$\dot{x}(t) = \int_s^t A(t, \tau) x(\tau)\, d\tau, \quad t > s, \qquad x(s) = 1, \tag{10}$$

where $A(t, \tau) = 0$ for $\tau > t$ and $A \in L^1([s, T] \times [s, T])$ for any T, with $s < T < \infty$.

Theorem 3. The solution of (10) is given by

$$\Phi(t, s) = 1 + \sum_{n=1}^{\infty} \int_s^t d\tau_1 \int_s^{\tau_1} d\tau_2 \cdots \int_s^{\tau_{2n-2}} d\tau_{2n-1}$$

$$\times \int_s^{\tau_{2n-1}} d\tau_{2n}\, A(\tau_1, \tau_2) \cdots A(\tau_{2n-1}, \tau_{2n}). \tag{11}$$

Let us consider the Volterra–Stieltjes integrodifferential equation

$$\dot{x}(t) = \int_s^t d_\tau\, A(t, \tau) x(\tau), \quad t > s, \qquad x(s) = 1, \tag{12}$$

where $A(t, \tau)$ is measurable in t, of bounded variation in τ for $s \le t, \tau \le T$, and constant for $\tau \ge t$, and

$$\int_s^T dt \left| \int_s^T d_\tau\, A(t, \tau) \right| < \infty, \qquad \text{for any} \quad s < T < \infty.$$

Theorem 4. The solution of (12) is given by

$$\Phi(t, s) = 1 + \sum_{n=1}^{\infty} \int_s^t d\tau_1 \int_s^{\tau_1} d_{\tau_2}\, A(\tau_1, \tau_2) \cdots$$

$$\times \int_s^{\tau_{2n-2}} d\tau_{2n-1} \int_s^{\tau_{2n-1}} d_{\tau_{2n}}\, A(\tau_{2n-1}, \tau_{2n}). \tag{13}$$

Theorems 1 to 4 hold for n-vector functional differential equations with appropriate modifications.

REFERENCES

[1] H. T. Banks, Representations of solutions of linear functional differential equations, *J. Differential Equations* **5** (1969), 399–409.
[2] M. C. Delfour and S. K. Mitter, Hereditary differential equations with constant delays II— A class of affine systems and the adjoint problem. *Rep.* CRM-293, Centre de Rech. Math., Univ. Montreal, Montreal, Canada.

Chapter 5: TOPOLOGICAL DYNAMICAL SYSTEMS

Extendability of an Elementary Dynamical System to an Abstract Local Dynamical System

TARO URA
Department of Mathematics, Faculty of Science
Kobe University, Kobe, Japan

1. Introduction

In the theory of differential equations, we first construct a family of small solution-curves, in assuming some continuity of the system of differential equations, and next proceed to prove that this family uniquely determines a family of large solution-curves nonextendable in the sense that each of the large solution-curves comes from and goes to the boundary of the domain of definition of the system. Proposing to himself the problem of clarifying the axiomatic foundation of this process, the author introduced a concept of an F-family of curves based on fundamental properties of small solution-curves of an autonomous system of differential equations and showed that an F-family determines a unique family of curves, denoted by **F**, satisfying the nonextendability condition [6]. Later Hájek introduced concepts of a (continuous) local dynamical system (generalization of a classical dynamical system) (cf. also [5]) and its germ [2-4] for the same purpose as the author's, the equivalence of which to F and **F** was intuitively clear and was strictly proved by Ahmad [1]. In the sequel, we shall use Hájek's axioms and terms, because they are easier to explain than the author's. However, we shall omit the qualifiers "dynamical" and "local dynamical" for simplicity of exposition.

Hájek's axioms are expressed in terms of the topological structure of the phase space and of the topological–algebraic structure of the topological abelian group of the real numbers. Thus our problem is considered as solved by topological–algebraic terms. On the other hand, to develop the study of continuous systems, it is natural to research purely algebraic properties of continuous systems apart from those depending on their topological structure. In the author's opinion, this is why Hájek introduced and studied the concept of an abstract system [2, 4]. Corresponding to the germ of a system, an elementary system was defined by the same author [4]. However, an elementary system does not play its role sufficiently for our purpose; more precisely, it determines *at most one* abstract system, but it may not determine *any* abstract system. Thus Hájek suggested the question of finding reasonable conditions on an elementary system for the existence of *at least one* such abstract system.

The purpose of this paper is to give a *complete* answer to this question, in other words, to solve our problem without use of topological properties of the phase space. The answer is the *no-intersection axiom* explained in Section 2, and is *complete* in the sense that an elementary system determines an abstract system, iff the former satisfies this axiom.

Because of the nature of our problem, our results obviously hold for a nonautonomous (and so a general) system of differential equations, if we consider the classical, so-called parametrized system corresponding to the given nonautonomous system. However, Theorem 1 in its form is not applicable to an abstract local *semidynamical* system, which is suggested by Theorem 2. Although we completely omit proofs, examples, and detailed explanations (which together with the discussion of related problems in wider categories will be published in [7]), we hope the main steps in the proof of Theorem 1 are made clear by the lemmas and propositions.

2. No-Intersection Axiom

Before explaining the no-intersection axiom, we shall recall some definitions and state some fundamental remarks. R will denote the set of real numbers with or without the usual topology.

Definition 1. Let X be an abstract set and D a subset of $X \times R$ expressed as

$$D = \bigcup_{x \in X} \{x\} \times D_x,$$

where for every $x \in X$, D_x is an open interval containing 0. A mapping $\pi: D \to X$ or the triple (X, D, π) is called an *elementary system on the phase set X with domain D*, iff

(AI)   For every $x \in X$, $\pi(x, 0) = x$;

(AII) If (x, t), $(x, t + s)$, $(\pi(x, t), s) \in D$, then $\pi(x, t + s) = \pi(\pi(x, t), s)$.

Definition 2. An elementary system (X, D, π) is called an *abstract system* iff it satisfies further:

(AII') (Nonextendability Axiom) If $(x, t) \in D$, then

$$D_x = D_{\pi(x, t)} + t.$$

Definition 3. Let X be a topological space. An elementary system (X, D, π) is called a *continuous germ on the phase space X* iff

(CO') D is a neighborhood of $X \times \{0\}$ in the product space $X \times R$ and $\pi: D \to X$ is continuous.

An abstract system (X, D, π) is called a *continuous system* iff it satisfies further:

(CO) (Openness Axiom) D is an open subset of $X \times R$ and $\pi: D \to X$ is continuous.

Remark. In the definition of a continuous system, the nonextendability axiom (AII') can be equivalently replaced by:

(CIII) (Nonextendability Axiom) For $x \in X$, put $D_x = (a_x, b_x)$. If b_x (or a_x) is finite, then the cluster set of $\pi(x, t)$ as $t \uparrow b_x$ (or $\downarrow a_x$) is empty.

The following is known:

(1) Let $(X, \mathscr{D}, \mu)$ be a continuous germ; then there exists a unique continuous system (X, D, π) such that $\pi|_{\mathscr{D}} = \mu$.

(2) Conversely, if a continuous system (X, D, π) is given, then for every neighborhood $\mathscr{D}$ of $X \times \{0\}$ contained in D expressed as $\mathscr{D} = \bigcup_{x \in X} \{x\} \times \mathscr{D}_x$, every $\mathscr{D}_x$ being an open interval containing 0, $(X, \mathscr{D}, \pi|_{\mathscr{D}})$ is a continuous germ.

Now we introduce the no-intersection axiom.

Definition 4. An elementary system $(X, \mathscr{D}, \mu)$ is called a *germ of an abstract system* (abbreviated as an *abstract germ*) iff it satisfies:

(AIII) (No-Intersection Axiom) If (x_1, t), $(x_2, t) \in \mathscr{D}$ and $\mu(x_1, t) = \mu(x_2, t)$, then $x_1 = x_2$.

Proposition 1. An abstract system is an abstract germ.

Lemma 1. Let $(X, \mathscr{D}, \mu)$ be an elementary system. Then for every pair $(x_1, t_1), (x_2, t_2) \in \mathscr{D}$, the set

$$A = \{s \in (\mathscr{D}_{x_1} - t_1) \cap (\mathscr{D}_{x_2} - t_2) | \mu(x_1, t_1 + s) = \mu(x_2, t_2 + s)\}$$

is open in R, and μ is an abstract germ iff every A is closed.

Proposition 2. Let $(X, \mathscr{D}, \mu)$ be an elementary system on a Hausdorff space X. If for every $x \in X$, $\mu(x, \cdot): \mathscr{D}_x \to X$ is continuous, then every A in Lemma 1 is closed, so that μ is an abstract germ.

Proposition 3. A continuous germ is an abstract germ.

(No separation assumption of the phase space is necessary.)

3. Main Theorems

Proposition 4. Let Φ be the set of all abstract germs on a fixed phase set X. Define a relation $\ll$ in Φ by

$$(X, \mathscr{D}^1, \mu^1) \ll (X, \mathscr{D}^2, \mu^2) \quad \text{or} \quad \mu^1 \ll \mu^2 \quad \text{iff} \quad \mathscr{D}^1 \subset \mathscr{D}^2, \quad \mu_1 = \mu_2|_{\mathscr{D}^1}.$$

Then the relation $\ll$ is a partial order in Φ.

Proposition 5. Let μ^0 be an abstract germ, and Φ^0 the set of all μ such that $\mu^0 \ll \mu$. Then, Φ^0 is a complete lattice.

Lemma 2. Every maximal element in Φ is an abstract system.

Theorem 1. For every abstract germ μ, we have a unique abstract system π (generated by μ) such that $\mu \ll \pi$.

Theorem 2. Let (X, D, π) be the abstract system generated by $(X, \mathscr{D}, \mu)$. For every $x_0 \in X$, one can fix a pair of chains in $\mathscr{D}$, $\{(x_k, \tau_k)\}_{k \in Z}$, $\{(x_k, \sigma_k)\}_{k \in Z}$, satisfying $\mu(x_{k-1}, \tau_{k-1}) = \mu(x_k, \sigma_k)$ such that for every $t \in D_{x_0}$, there exist $k \in Z$ and $\tau \in \mathscr{D}_{x_k}$ that satisfy $\pi(x_0, t) = \mu(x_k, \tau)$. Here Z denotes the set of integers.

Compare this to the following known theorem: If π and μ in Theorem 2 are continuous, then for every $x_0 \in X$, one can fix a chain in $\mathscr{D}$, $\{(x_k, \tau_k)\}_{k \in Z}$ satisfying $\mu(x_{k-1}, \tau_{k-1}) = x_k$, such that for every $t \in D_{x_0}$, there exist $k \in Z$ and $\tau \in \mathscr{D}_{x_k}$ that satisfy $\pi(x_0, t) = \mu(x_k, \tau)$.

One can verify by an example that Theorem 2 can not be strengthened as the theorem recalled here.

REFERENCES

[1] S. Ahmad, On Ura's axioms and local dynamical systems, *Funkcial. Ekvac.* **12** (1969), 181–191.

[2] O. Hájek, Structure of dynamical systems, *Comment. Math. Univ. Carolinae* **6** (1965), 53–72.

[3] O. Hájek, Local characterization of local semidynamical systems, *Math. Syst. Theory* **2** (1968), 17–25.

[4] O. Hájek, "Dynamical Systems in the Plane." Academic Press, New York, 1968.

[5] G. R. Sell, Nonautonomous differential equations and topological dynamics I, *Trans. Amer. Math. Soc.* **127** (1967), 241–262.

[6] T. Ura, Sur le courant extérieur à une région invariante; prolongement d'une caractéristique et l'ordre de stabilité, *Funkcial. Ekvac.* **2** (1959), 143–200.

[7] T. Ura, Local determinancy of abstract local dynamical systems, *Math. Syst. Theory* **9** (1975), 159–189.

Skew-Product Dynamical Systems*

ROBERT J. SACKER

Department of Mathematics
University of Southern California, Los Angeles, California

1. Introduction

The skew-product flow is the appropriate setting for studying many of the qualitative properties of nonautonomous ordinary differential equations, functional differential equations, finite difference equations, and mappings of manifolds. Its appropriateness derives from the fact that many important questions can be treated within the framework of the dynamics of compact spaces while retaining a setting in which one's geometric intuition can freely play a role. The idea of the skew-product flow appeared earlier in the paper of Miller [4].

Abstractly, a skew-product flow consists of a pair of flows

$$\pi: X \times Y \times \mathscr{T} \to X \times Y, \qquad \text{and} \qquad \sigma: Y \times \mathscr{T} \to Y,$$

where the natural projection $p: X \times Y \to Y$ commutes with the action of the flows. Here $\mathscr{T}$ is a topological group (usually the reals $\mathbb{R}$ or the integers Z). The commutativity means that

$$\pi(x, y, t) = (\varphi(x, y, t), \sigma(y, t)).$$

For ordinary differential equations, we consider the space $\mathbb{R}^n \times Y$ (or $\mathbb{C}^n \times Y$), where Y is an appropriately topologized space of functions $f = f(t, x)$, i.e., $f: R \times R^n \to R^n$, which is invariant with respect to the flow $\sigma(f, \tau) = f_\tau$, where $f_\tau(t, x) = f(t + \tau, x)$. On the product we put the flow $\pi(\xi, f, \tau) = (\varphi(\xi, f, \tau), f_\tau)$, where $\varphi(\xi, f, \tau)$ is the solution, at time τ, of the initial value problem

$$\dot{x} = f(t, x), \qquad x(0) = \xi.$$

We assume that f satisfies conditions guaranteeing existence, uniqueness, and all solutions under consideration can be continued to the whole real axis. Other possibilities are discussed in the paper of Miller and Sell [5].

---

* This research was supported in part by U.S. Army Contract DAHC04-74-G-0013.

For functional differential equations of retarded type we consider the product space $C \times Y$, where C is the Banach space of continuous functions from $[-r, 0]$ to R^n with the sup norm and Y a space of functionals $f: R \times C \to R^n$. As above, the flow σ on Y is given by $\sigma(f, \tau) = f_\tau$ but now $\pi(v, f, \tau) = (\varphi(v, f, \tau), f_\tau)$, where $\varphi(v, f, \tau)$ is the $(\tau - r, \tau)$-profile of the solution of $\dot{x} = f(t, x_t)$ satisfying the initial condition $x_0 = v$. In the notation of Hale [2; p. 33], $\varphi(v, f, \tau) = T(\tau, 0)v$. Even when f satisfies conditions guaranteeing uniqueness and continuation to the positive reals of a bounded solution, it may still be the case that such a solution cannot be extended to the negative reals. Thus, in general, π is just a semiflow that projects to a flow σ.

For finite difference equations

$$x_{j+1} - x_j = f(j, x_j), \qquad x_j \in R^n,$$

we consider the product space $R^n \times Y$, where Y is a space of functions $f: Z \times R^n \to R^n$ and the flow σ on Y is given by the same formulas as before. In the expression for π, $\varphi(x_0, f, \tau)$ is obtained by iterating the difference equation τ times with initial condition x_0. If the equation cannot be iterated in the backward direction, then as in the case of functional differential equations we obtain a semiflow that projects to a flow.

In discussing mappings of manifolds one must replace the skew-product flow by its generalization, the fiber-preserving flow on a vector bundle. If $F: Y \to Y$ is a diffeomorphism of a smooth compact manifold Y, then one obtains a flow $\sigma: Y \times Z \to Y$ on Y by setting $\sigma(y, n) = F^n(y)$, the nth iterate of F applied to y (or the $-n$th iterate of F^{-1} if $n < 0$). If $x \in Y_y$, the tangent space of Y at y, then we define the flow π on the tangent bundle TY by setting $\pi(x, y, n) = (DF^n(y)x, F^n(y))$, where $DF^n(y)$ is the derivative of F^n at y.

It is shown by Sacker and Sell [7, 8] that if

(1)   for each $y \in Y$ the vector $x = 0$ is the only vector for which $\|DF^n(y)x\|$ is bounded uniformly for $n \in Z$, and

(2)   the collection of minimal (compact) subsets of Y is dense in Y,

then F is an Anosov diffeomorphism, i.e., the tangent bundle admits a global invariant splitting $TM = \mathscr{S} + \mathscr{U}$ (Whitney sum) into the stable and unstable bundles and the rates of decay are exponential. Diffeomorphisms satisfying (1) are called quasi-Anosov by Mañé [3] (see also Robinson [6]). Recently Sacker and Sell [9] have shown that hypothesis (2) can be replaced by the assumption that the flow σ on Y is chain-recurrent. For parallel results along these lines see Selgrade [10]. The concept of chain-recurrence is due to Conely and others (its history is uncertain) and is discussed in [1].

For a discussion of integral equations and local semiflows see [5]. Also see [5] for further discussion of skew-product flows and some recent results of Sacker and Sell concerning existence of almost-periodic solutions of nonlinear almost-periodic ordinary differential equations and the existence of dichotomies and invariant splittings, together with a related spectral theory for linear differential systems. See also [12].

2. Bounded Solutions of Linear Systems

In this section we will present some very recent results of Sacker and Sell [9] giving estimates on the number of independent bounded orbits for linear skew-product systems. We will content ourselves with specific cases of more general theorems given in [9] and [13].

Let $\mathscr{B}$ denote the linear subspace of R^n consisting of initial conditions giving rise to solutions of

$$\dot{x} = A(t)x, \qquad x \in R^n,$$

that are bounded on the entire t axis. Here A is assumed to lie in some translation invariant space Y of $n \times n$ matrix-valued functions defined on R such that the mapping $\sigma(A, \tau) = A_\tau$ is a flow on Y [as usual, $A_\tau(t) = A(t + \tau)$]. For definiteness let us consider the space Y of all $n \times n$ matrix-valued functions that are continuous with the topology of uniform convergence on compact subsets of R. For other examples see [8, p. 439] and the lectures of Sell [11, Chap. 3].

Theorem 1. Assume $\lim_{t \to \pm\infty} A(t) = A_\pm$, where $A_\pm$ have no purely imaginary eigenvalues. Let d_+ and d_- denote the number of eigenvalues of A_+ and A_-, respectively, having negative real parts. Then

$$\dim \mathscr{B} \geq d_+ - d_-.$$

Moreover, the norm of each solution $\varphi(t)$ having initial conditions in $\mathscr{B}$ approaches zero as $|t| \to \infty$.

The next theorem is more general and does not require the evaluation or even the existence of the limits of A.

Theorem 2. Let A_+^* and A_-^* be constant $n \times n$ matrices having no purely imaginary eigenvalues and let d_+^* and d_-^* denote respectively the number of eigenvalues having negative real parts. Then there exists an $\varepsilon > 0$ with the following property: Suppose there exists a $T > 0$ such that

$$\|A(t) - A_+^*\| < \varepsilon \qquad \text{for all} \quad t \geq T,$$

$$\|A(t) - A_-^*\| < \varepsilon \qquad \text{for all} \quad t \leq -T.$$

Then the same conclusions of Theorem 1 hold.

For the linear difference equation

$$x_{j+1} - x_j = C(j)x_j,$$

the above theorems hold with the following modifications: $A(j) = I + C(j)$, the condition "no purely imaginary eigenvalues," must read "no eigenvalues of unit modulus," and $d_\pm$ should denote the number of eigenvalues of $A_\pm$ having modulus < 1. The variable t is of course now an integer and $\mathcal{B}$ denotes the subspace of values x_0 such that the sequence $\{x_j\}_{j=-\infty}^{+\infty}$ is bounded.

When the equation arises as the finite difference approximation with uniform net spacing

$$[x(t_{j+1}) - x(t_j)]/h = P(t_j)x(t_j)$$

of the ODE $\dot{x} = P(t)x$ with bounded P, then for $0 < h < (\sup\|P(t)\|)^{-1}$, A is invertible and the difference equation can be iterated in the backward direction thus defining a flow.

REFERENCES

[1] C. C. Conley, Some aspects of the qualitative theory of differential equations, Volume 1, Chapter 1 of this treatise.

[2] J. K. Hale, "Functional Differential Equations." Springer, New York, 1971.

[3] R. Mañé, Persistent manifolds are normally hyperbolic, *Bull. Amer. Math. Soc.* **80** (1974), 90–91.

[4] R. K. Miller, Almost periodic differential equations as dynamical systems with applications to the existence of a.p. solutions, *J. Differential Equations* **1** (1965), 337–345.

[5] R. K. Miller and G. R. Sell, Topological dynamics and its relation to integral equations and nonautonomous systems, Volume 1, Chapter 5 of this treatise.

[6] R. Clark Robinson, A quasi Anosov flow that is not Anosov, *Indiana J. Math.* (to appear).

[7] R. J. Sacker and G. R. Sell, A note on Anosov diffeomorphisms, *Bull. Amer. Math. Soc.* **80** (1974), 278–280.

[8] R. J. Sacker and G. R. Sell, Existence of dichotomies and invariant splittings for linear differential systems I, *J. Differential Equations* **15** (1974), 429–458.

[9] R. J. Sacker and G. R. Sell, Part II of [8], *J. Differential Equations* (to appear).

[10] James Selgrade, Isolated invariant sets for flows on vector bundles, *Trans. Amer. Math. Soc.* **203** (1975) 359–390.

[11] G. R. Sell, "Lectures on Topological Dynamics and Differential Equations," Van-Nostrand-Reinhold, Princeton, New Jersey, 1971.

[12] R. J. Sacker, Linear skew-product dynamical systems, *Proc. 3rd Mexico–U.S.A. Symp. Differential Equations* (to appear).

[13] R. J. Sacker and G. R. Sell, Part IV of [8], *J. Differential Equations* (to appear).

Liapunov Functions and the Comparison Principle

PETER SEIBERT *
Instituto de Matemática
Universidad Católica de Chile, Santiago, Chile

Introduction

In this paper we present a general principle that contains, in the most universal form possible, the basic idea underlying the notion of a Liapunov function (abbreviated L-function). The objects of our theory are abstract systems consisting of a set (the "state" or "event space") endowed with a binary relation, usually a preorder (or quasi-order, the "flow"), and two collections of subsets, called "quasifilters" (generalized neighborhood filters, corresponding to initial and subsequent states of a motion, respectively). The concept of Liapunov stability can be extended to this class of systems in a natural way. In the spirit of the comparison principle for differential equations, we consider pairs of such systems together with mappings from one state space to the other. Assuming stability of one of the two systems (serving as the comparison system), we formulate sufficient conditions for stability of the other. These conditions still show a strong similarity to Zubov's conditions for Liapunov stability in the context of dynamical systems in metric spaces; we therefore call the functions satisfying them "Liapunov mappings."

If the state space of the comparison system is linearly ordered (in particular, if it is a subset of the real line), we obtain a class of generalized (scalar) L-functions, in which the condition of monotonicity is replaced by a weaker one, which amounts to stability of a scalar comparison system. Functions of this type were called "para-Liapunov" by Hájek [2]. Yorke proved [8], for the case of an autonomous differential equation, that stability of the origin implies the existence of a continuous para-L-function, while under the same conditions a continuous L-function (not depending explicitly on the time) does not exist necessarily. The theory of L-functions at this level of generality was presented in [5–7].

* Present address: Departmento de Matemáticas, Universidad Simón Bolívar, Sartenejas, Baruta Edo Miranda, Venezuela.

Here we sketch only the basic part of the theory, limiting ourselves to the statement of sufficient conditions for stability, while the question of inverse theorems, in the case of a scalar L-functions, is treated by Salzberg in Chapter 5 of this volume [4]. The general problem of existence of L-mappings is at present completely open.

By appropriate specializations and adaptations, our theory can be applied to a wide range of special cases and related problems. For instance, by a mere change of the formalism, one obtains results on "weak stability" in the sense of Roxin [3], which may be interpreted as controllability. This facet of the theory is sketched in Section 4. By defining the system in a product space with the real line as one factor ("state-time" or "event space"), one can retrieve the classical theorems on Liapunov stability in nonautonomous systems using time-dependent L-functions, including such variants as eventual stability (cf. [6]). Finally, by a suitable adaptation of the main theorem on boundedness (the "dual" concept to stability), one gains an alternative approach to Liapunov-type theorems on attraction.

1. General Definitions and Notations

We will denote by X a set (the "state" or "event space"), by $\mathscr{X}$ the collection of all nonvoid subsets of X, and by Φ a binary relation on X (the "flow"). A *quasifilter* on X is a nonvoid subset of $\mathscr{X}$. Throughout, binary relations will be denoted by capital Greek letters and quasifilters by capital script letters. If $\mathscr{A}$ and $\mathscr{B}$ are two quasifilters on the same set, we say "$\mathscr{A}$ is *coarser* than $\mathscr{B}$," in symbols $\mathscr{A} \dashv \mathscr{B}$, if every element of $\mathscr{A}$ contains an element of $\mathscr{B}$.

We give some notational conventions: If f is a (possibly set-valued) function defined on X, we also denote by f the function defined on $\mathscr{X}$ by

$$f(A) = \bigcup \{f(x) | x \in A\}, \qquad A \in \mathscr{A}.$$

If $\mathscr{A}$ is a quasifilter on the domain of the function f, we define the quasifilter $f\mathscr{A}$ as $\{fA | A \in \mathscr{A}\}$.

Moreover, in order to avoid a proliferation of parentheses indicating variables of functions, we will usually omit these. For a similar reason, we will write compositions of functions without connecting symbols.

Finally, instead of $(x, y) \in \Phi$, we will write $y \in \Phi(x)$, or $y \in \Phi x$.

By a *system* we mean a quadruplet $S_X = (X, \Phi, \mathscr{D}, \mathscr{E})$, where X denotes a set, Φ a binary relation on X, and $\mathscr{D}$ and $\mathscr{E}$ quasifilters on X.

The system $S_X = (X, \Phi, \mathscr{D}, \mathscr{E})$ is called *stable* if $\mathscr{E} \dashrightarrow \Phi\mathscr{D}$; explicitly, if every element of $\mathscr{E}$ contains the set ΦD for some $D \in \mathscr{D}$.

Example 1. If X is a topological space, Φ the positive semiorbit relation of a dynamical system on X, $M \subset X$ compact, and $\mathscr{N}_M$ denotes the neighborhood filter of M, then stability of $(X, \Phi, \mathscr{N}_M, \mathscr{N}_M)$ is Liapunov stability of M.

Example 2. If X is a metric space, $M \subset X$ closed, Φ and $\mathscr{N}_M$ as in Example 1, and $\mathscr{M}_M$ denotes the quasifilter of ε-neighborhoods of M $(\varepsilon > 0)$, then $(X, \Phi, \mathscr{N}_M, \mathscr{M}_M)$ is "stability" in the sense of Bhatia and Szegö [1, Chap. V], a concept modeled on nonuniform stability of the origin in a nonautonomous system.

2. Liapunov Mappings

We now consider, simultaneously, two systems

$$S_X = (X, \Phi, \mathscr{D}, \mathscr{E}), \quad \text{and} \quad S_Y = (Y, \Psi, \mathscr{F}, \mathscr{G}).$$

The idea is this: One supposes S_Y is known to be stable, and asks for conditions guaranteeing stability of the system in question, S_X. To this end we consider mappings v from X into Y and define $v: X \to Y$ as a *Liapunov mapping* from S_X to S_Y if it satisfies the following conditions:

$$v^{-1}\mathscr{F} \text{ and } v^{-1}\mathscr{G} \text{ are quasifilters;} \tag{2.0}$$

$$\mathscr{E} \dashrightarrow v^{-1}\mathscr{G} \tag{2.1}$$

$$v^{-1}\mathscr{F} \dashrightarrow \mathscr{D}, \tag{2.2}$$

$$v^{-1}\Psi\mathscr{F} \dashrightarrow \Phi v^{-1}\mathscr{F}. \tag{2.3}$$

Theorem 1. If there exists a Liapunov mapping from the system S_X to a stable system S_Y, then S_X is also stable.

3. Para-Liapunov Functions

We now focus our attention on the important special case where Y is the real line R, Ψ is the relation $\leq$, and $\mathscr{G}$ coincides with the quasifilter $\mathscr{F}$ defined by

$$\mathscr{F} = \{F_\beta^< \,|\, \beta \in R \text{ such that } F_\beta^< \neq \varnothing\}, \tag{A}$$

where

$$F_\beta^< = \{\alpha \in v(X) \,|\, \alpha < \beta\}. \tag{B}$$

To simplify the notation, we put $v^{-1}\mathscr{F} = \mathscr{S}_v$; this is the quasifilter consisting of the (nonvoid) sets on which v is smaller than a given value. In this case conditions (2.0)–(2.3) reduce to the following:

$$v \text{ is not a constant,} \tag{3.0}$$

$$\mathscr{E} \prec \mathscr{S}_v, \tag{3.1}$$

$$\mathscr{S}_v \prec \mathscr{D}, \tag{3.2}$$

$$\mathscr{S}_v \prec \Phi\mathscr{S}_v. \tag{3.3}$$

We call a function from X into R, with these properties, a *para-L-function* with respect to $(X, \Phi, \mathscr{D}, \mathscr{E})$.

Corollary 1. The existence of a para-L-function with respect to the system $S_X = (X, \Phi, \mathscr{D}, \mathscr{E})$ implies the stability of S_X.

Remark. Condition (3.3) obviously holds if v is nonincreasing under Φ, i.e., if $y \in \Phi(x)$ implies $v(y) \le v(x)$. Under this stronger condition and with (3.0)–(3.2) as before, we call v a *Liapunov function (L-function)*.

A necessary and sufficient condition for the existence of an L-function is given by Salzberg and the author in [5] (summarized in [4]).

4. Weak Stability or Controllability

In his studies of stability in dynamical polysystems [3], Roxin considered, besides the usual concept of stability, also a weak form that may be interpreted as controllability if the system is considered as a control system.

Given a preordered set (X, Φ), we say a subset $\gamma \subset X$ is a *semiorbit starting at* x_0 if it is a maximal set with the properties that $x \in \Phi(x_0)$, for every $x \in \gamma$, and γ is totally preordered by Φ. The set of all semiorbits starting at x will be denoted by $\Gamma(x)$.

Supposing $\mathscr{D}$ and $\mathscr{E}$ are quasifilters on X, we say the system $(X, \Phi, \mathscr{D}, \mathscr{E})$ is *controllable* if for any $E \in \mathscr{E}$ there exists a $D \in \mathscr{D}$ such that for each $x \in D$ there exists a $\gamma \in \Gamma(x)$ contained in E.

We introduce a *selection function* σ from sets $A \subset X$ into ΓA, which assign to every $x \in A$ a semiorbit $\gamma_\sigma(x) \in \Gamma(x)$, and denote by ΣA the set of all σ's with domain A. Next, we define for every set $A \in \mathscr{X}$ and every $\sigma \in \Sigma A$ the set $\Phi_\sigma = \bigcup \{\gamma_\sigma(x) \mid x \in A\}$, and for every quasifilter $\mathscr{A}$ in X the quasifilter $\Phi_c \mathscr{A} = \{\Phi_\sigma A \mid \sigma \in \Sigma A, A \in \mathscr{A}\}$, which is the collection of all sets attainable from sets of $\mathscr{A}$ by some selection.

Now the definition of controllability may be formulated more concisely by the condition $\mathscr{E} \rightarrowtail \Phi_c \mathscr{D}$.

We call a function $v: X \to Y$ a *weak L-mapping* from S_X to S_Y (defined as in Section 2) if it satisfies conditions (2.0)–(2.3), with Φ replaced by Φ_σ and Ψ by Ψ_c (defined analogously) in (2.3).

Theorem 2. If there exists a weak L-mapping from S_X to a controllable system S_Y, then S_X is also controllable.

The same specialization as in Section 3 yields a result analogous to corollary 1. The analog of condition (3.3) is satisfied, in particular, if every point of X is the starting point of a semiorbit along which v is nonincreasing.

REFERENCES

[1] Bhatia, N. P., and Szegö, G. P. "Stability Theory in Dynamical Systems." Springer, New York, 1970.

[2] Hájek, O., Ordinary and asymptotic stability of noncompact sets, *J. Differential Equations* **11** (1972), 49–65.

[3] Roxin, E., Stability in general control systems, *J. Differential Equations* **1** (1965), 115–150. 115–150.

[4] Salzberg, P., Existence and continuity of Liapunov functions in general systems, Chapter 5 in this volume.

[5] Salzberg, P. M., and Seibert, P., A necessary and sufficient condition for the existence of a Liapunov function, *Funkcial. Ekvac.* **16** (1973), 97–101.

[6] Seibert, P., A general theory of Liapunov stability, *Ord. Differential Equations NRL-MRC Conf., Washington, D.C., 1971*, pp. 563–568. Academic Press, New York, 1972.

[7] Seibert, P. V., A unified theory of Liapunov stability, *Funkcial. Ekvac.* **15** (1972), 139–147.

[8] Yorke, J. A., Invariance for ordinary differential equations, *Math. Syst. Theory* **1** (1967), 353–372.

Distal Semidynamical Systems

NAM P. BHATIA and M. NISHIHAMA
Division of Mathematics and Physics
University of Maryland, Baltimore, Maryland

R. Ellis has studied the so-called distal transformation groups [1, 2]. In [1] he introduced the closure of the set of transitions and proved the same to be a semigroup, the so-called enveloping semigroup [2]. If all trajectory closures are compact, then a transformation group is distal if and only if its enveloping semigroup is a group ([1, Theorem 1] or [3]).

In the case of dynamical systems, i.e., transformation groups where the action is via the reals or the integers, one can introduce the notions of positively (and negatively) distal dynamical systems, as is the case with many other notions. However, one can show that positively distal dynamical systems are distal whenever all positive trajectory closures are compact. It is therefore obvious that almost all known results on distal flows [1–3] are available for positively distal dynamical systems.

In the case of semidynamical systems [4–6] where only forward uniqueness is assumed and generally there is nonuniqueness in the backward direction, logically only the notion of positive distality can be introduced. It is also not obvious as to which of the known results on distal flows may be carried over to this case. Here we study (positively) distal semidynamical systems and show that major results on distal flows are carried over to this case. We actually show that (positively) Lagrange stable distal semidynamical systems are indeed dynamical. This circumstance results in the generalization of a result of Bhatia and Chow [4], namely, that any positively almost-periodic point of a semidynamical system on a complete metric space has a compact positive trajectory closure and the restriction of the semisystem to this set is dynamical. Finally, we develop an analytic characterization of distal semidynamical systems that appears to be a completely new result even for dynamical systems.

1. Notation, Definitions, and Some Known Results

X denotes a Hausdorff space, G^+ denotes the nonnegative reals or the nonnegative integers, and G denotes the reals or the integers.

A semidynamical system on X [4–5] is a continuous map π from $X \times G^+$ into X satisfying $x\pi 0 = x$ and $(x\pi t)\pi s = x\pi(t + s)$ for all $x \in X$ and t, s in G^+. A dynamical system on X is defined by replacing G^+ by G in the foregoing statement.

We assume given a semidynamical system on X.

Definition 1.1. π is positively distal (or simply distal) if $(x\pi t_i \to z \leftarrow y\pi t_i)$ implies $x = y$ for any net t_i in G^+ and x, y, z in X.

Definition 1.2. π is positively Lagrange stable (or simply Lagrange stable) if the closure of every positive trajectory $x\pi G^+$ is compact. Here $x\pi G^+ = \{x\pi t : t \in G^+\}$ for any $x \in X$.

Definition 1.3. [4].   Let X be metric with distance d. A point x in X is said to be positively almost periodic if for every $\varepsilon > 0$ there is a relatively dense subset $D \subset G^+$ such that $d(x\pi t, x\pi(t + \tau)) < \varepsilon$ for all t in G^+ and τ in D.

Definition 1.4. A set M is positively invariant if $M = \bigcup\{x\pi G^+ : x \in M\}$. It is called positively minimal if it is closed and positively invariant and contains no proper nonempty closed positively invariant subset. We say π is pointwise minimal if every positive trajectory closure is positively minimal.

X^X denotes the topological space of all transformations of X into X endowed with the topology of pointwise convergence. Given a semidynamical system π on X and $t \in G^+$, the transition π_t from X into X given by $x\pi_t = \pi_t(x) = x\pi t$ for every $x \in X$ is continuous. The set of transitions $T = \{\pi_t : t \in G^+\}$ and X are both semigroups under the operation of composition of maps, and the transition π_0 is an identity for both. The closure of T in X^X is denoted by E and following Ellis one can show that E is a semigroup too, the so-called enveloping semigroup.

2. The Main Results

As proved for transformation groups by Ellis we have

Lemma 2.1. If π is Lagrange stable and distal, then the closure E of T (i.e., the enveloping semigroup) is a compact group. Moreover, the maps $f_p : E \to E$ given by $f_p(q) = p \circ q$ are continuous for each p in E.

Lemma 2.2. If π is Lagrange stable and distal, then every positive trajectory closure is positively minimal.

Theorem 2.3. Let π be Lagrange stable. Then the following are equivalent:

(2.3.1) E is a compact group.
(2.3.2) For every cardinal $a > 0$, the semidynamical system π^a on X^a is pointwise minimal.
(2.3.3) The semidynamical system π^2 on X^2 is pointwise minimal.

In addition to carrying over Ellis's results to the case of semidynamical systems we have

Theorem 2.4. If π is Lagrange stable and distal, then there is a unique abstract dynamical system π^* on X (i.e., π^* satisfies all but the continuity hypothesis of a dynamical system) that is dynamical on every compact invariant set. Moreover, π^* is dynamical on X if X is locally compact.

Theorem 2.5. Let X be complete metric. If every $x \in X$ is positively almost periodic, then π is Lagrange stable and distal.

And finally we have the following characterization:

Theorem 2.6. Let π be Lagrange stable. Then π is distal if and only if for every net t_i in G^+, one has $X = \{z \in X : x\pi t_j \to z$ for some $x \in X$ and some subnet t_j of $t_i\}$.

The proofs of Lemmas 2.1 and 2.2 and Theorem 2.3 may be constructed as in [1], and those of Theorems 2.4 and 2.5 as in [4]. We summarize the proof of Theorem 2.6.

Proof of Theorem 2.6. Let π be distal. Then E is a compact group. Let t_i be any net in G^+. Since π_{t_i} is in T and E is compact, there is a subnet t_j such that $\pi_{t_j} \to p \in E$. Since E is a group, there exists p^{-1} in E. Let $z \in X$. Set $x = zp^{-1}$, so that $z = xp$. This shows that $x\pi_{t_j} = x\pi t_j \to xp = z$. The "if" part is proved. For the "only if" part, notice first that Lagrange stability implies that the enveloping semigroup E is compact. Moreover, the maps f_p from E to E given by $f_p(q) = p \circ q$ are continuous. Using Lemma 1 in [1], there is an idempotent u in $pE = \{p \circ q : q \in E\}$ for any $p \in E$. Thus there is a q in E such that $p \circ q = u$. There is then a net t_i in G^+ such that $\pi_{t_i} \to u$. Let $z \in X$. By assumption there is a subnet t_j of t_i and an $x \in X$ such that $x\pi_{t_i} = x\pi t_j \to z$. This implies that $xu = z$. Consequently, $zu = (xu)u = x(u^2) = xu = z$. Thus u is an identity transformation in E. We have in fact shown that E is a group. Hence π is distal.

The following example shows that a Lagrange stable distal semidynamical system on an arbitrary metric space need not be dynamical.

Example 2.7. Consider the subset X of the euclidean plane consisting of the points of the open unit disc and the point $(1, 0)$. The following differential system (polar coordinates) defines a Langrange stable distal semidynamical system on X, which is not dynamical:

$$\dot{r} = 0, \qquad \dot{\theta} = \begin{cases} 1 & \text{if } \pi \leq \theta < 2\pi, \\ 1 - r & \text{if } 0 \leq \theta < \pi. \end{cases}$$

Note that the right-hand side is necessarily discontinuous.

ACKNOWLEDGMENT

The research of the first author is supported partially by the National Science Foundation through Grant NSF-GP-39057. The second author wishes to thank Dr. Morton Baratz, Vice-Chancellor, UMBC, for support of his research.

REFERENCES

[1] R. Ellis, Distal transformation groups, *Pacific J. Math.* **8** (1958), 401–405.
[2] R. Ellis, "Lectures on Topological Dynamics." Benjamin, New York, 1969.
[3] H. Furstenberg, Structure of distal flows, *Amer. J. Math.* **85** (1963), 477–515.
[4] N. P. Bhatia and S. N. Chow, Weak attraction, minimality, recurrence, and almost periodicity in semisystems, *Funkcial. Ekvac.* **15** (1972), 39–59.
[5] N. P. Bhatia and O. Hajek, Local semidynamical Systems, *Lecture Notes Math.* **90**, Springer-Verlag, Berlin and New York, 1969.
[6] J. Hale, "Functional Differential Equations" (Appl. Math. Ser. No. 3). Springer-Verlag, Berlin and New York, 1971.

Prolongations in Semidynamical Systems

PREM N. BAJAJ
Department of Mathematics
Wichita State University, Wichita, Kansas

Introduction

Semidynamical systems (sds) are continuous flows defined for all future time. Natural examples of sds are furnished by functional differential equations for which existence and uniqueness conditions hold. Though a substantial part of the theory of dynamical systems (ds) extends to sds, many new and interesting notions (e.g., a start point and a singular point [2–4]) arise in sds.

Prolongations in ds have been studied by numerous authors (Auslander and Seibert [1], Hajek [6–7], Ura [11]) and are useful in stability problems. However, prolongations, when defined as in ds, lose some of their basic properties. In this paper, we define another system of prolongations, examine which of the lost properties are restored, and state some results on prolongations and their limit sets.

Preliminaries

An sds is a pair (X, π) where X is a Hausdorff topological space, and $\pi: X \times R^+ \to X$ is a continuous map satisfying the following conditions:

(a) $\pi(x, 0) = x$, $x \in X$ (identity axiom).

(b) $\pi(\pi(x, t), s) = \pi(x, t + s)$, $x \in X$, $t, s \in R^+$ (semigroup axiom).

(R^+ denotes the set of nonnegative reals with the usual topology.) For brevity, $\pi(x, t)$ will be denoted by xt. If M is a subset of X and K a subset of R^+, then the set $\{xt : x \in M \subset X, t \in K \subset R^+\}$ will be denoted by MK. Positive trajectory $\gamma^+(x)$ and positive invariance are defined as in dynamical systems [1, 5]. A point x in X is said to be (positively) critical if $\gamma^+(x) = \{x\}$.

Throughout this paper, (X, π) denotes a semidynamical system. A net in X will be denoted by x_i, where i is an element of the directed set.

For any x in X, define positive limit set $\Lambda(x)$, positive prolongation $D(x)$, and positive prolongational limit set $J(x)$ as follows:

$\Lambda(x) = \{y \in X :$ there exists a net t_i in R^+, $t_i \to \infty$ such that $xt_i \to y\}$.

$D(x) = \{y \in X :$ there exists a net x_i in X, $x_i \to x$, and a net t_i in R^+ such that $x_i t_i \to y\}$.

$J(x) = \{y \in X :$ there exists a net x_i in X, $x_i \to x$, and a net t_i in R^+, $t_i \to \infty$ such that $x_i t_i \to y\}$.

Remark.   The definitions of $D(x)$, $J(x)$ being the same as in dynamical systems, the observation that sds are continuous maps defined for all non-negative t (i.e., for future time) tempts one to believe that their properties in dynamical systems will also hold in sds. This, however, is not true as the following illustration shows.

Example.   The equations

$$\frac{dr}{dt} = \frac{r}{1+r}, \qquad \frac{d\theta}{dt} = \frac{1}{1+r},$$

(in polar coordinates) represent for $r \neq 0$, a family of equiangular spirals, $r = 0$ being a critical point.

On using the transformation $x = u/(1 - u^2)$, $y = v$, where $|u| < 1$, du/dt and dv/dt are easily obtained [9, p. 343]. Now we complete our space by addition of the lines $u = \pm 1$, and $\{(u, 0) : u < -1\}$ with the limiting equations:

$$\frac{du}{dt} = \begin{cases} 0 & \text{for} \quad u = \pm 1, \\ 1 & \text{for} \quad u < -1, \end{cases}$$

$$\frac{dv}{dt} = \begin{cases} 1 & \text{for} \quad u = 1, \\ -1 & \text{for} \quad u = -1, \\ 0 & \text{for} \quad u < -1. \end{cases}$$

This defines a semidynamical system.

Let p be the point $(-2, 0)$, and q the point $(-1, -1)$ in the u, v plane. Then $q = p.2$, $J(p)$ is the empty set, $J(q)$ is the set $\{(u, v) : u = 1\}$ so that $J(pt) \neq J(p)$ for $t \geq 1$. For $t \geq 1$, even $D(pt) \subset D(p)$ does not hold. Examples can be constructed to show that even

$$J(x)t = J(x), \quad t \in R^+, \quad \text{and} \quad J(x) = \bigcap \{D(xt) : t \in R^+\},$$

do not, in general, hold in sds. Thus we are led to the following:

Definitions. Let (X, π) be an sds. Define maps B and L on X with values in the set of subsets of X by

$B(x) = \{y \in X : \text{there exists a net } x_i \text{ in } X, x_i \to xs, \text{ for some } s \text{ in } R^+,$
      $\text{and a net } t_i \text{ in } R^+ \text{ such that } x_i t_i \to y\},$

$L(x) = \{y \in X : \text{there exists a net } x_i \text{ in } X, x_i \to xs, \text{ for some } s \in R^+,$
      $\text{and a net } t_i \text{ in } R^+, t_i \to +\infty \text{ such that } x_i t_i \to y\}.$

It is clear from the definition that

(a) $B(x) = \bigcup\{D(xt) : t \in R^+\} \supset D(x),$
(b) $L(x) = \bigcup\{J(xt) : t \in R^+\} \supset J(x),$
(c) $B(xt) \subset B(x)$ whenever $t \in R^+.$

Theorem. Let (X, π) be a semidynamical system. Let $x \in X$, $t \in R^+$. Then $L(xt) = L(x)$.

Proof. Notice that $J(x) \subset J(xt)$ whenever $t \in R^+.$

Theorem. Let (X, π) be a semidynamical system. Then

(a) $B(x) = L(x) \cup \gamma^+(x),$
(b) $\mathrm{Cl}(B(x)) = \mathrm{Cl}(L(x)) \cup \gamma^+(x).$

In general, $B(x)$ and $L(x)$ are not closed. The above theorem, however, states that set theoretic maps B and L, and also their closures, satisfy a relation similar to $D(x) = J(x) \cup \gamma^+(x)$.

Theorem. Preserving the notation,

$$L(x) = \bigcap\{B(xt) : t \in R^+\}.$$

Remark. Out of the properties stated in the example to have been lost by the usual prolongations, only $L(x)t = L(x)$ [the analog of $J(x)t = J(x)$] remains to be examined. In sds, $L(x)t \neq L(x)$ in general. Due to the points with finite escape time, for any subset K of X, we cannot expect $Kt = K$ for any $t \in R^+$, even if K is positively invariant.

Theorem. Let (X, π) be a semidynamical system. Let X be rim compact [8, 10]. Let $x \in X$. If $L(x)$ is relatively compact, then $\Lambda(x)$ is non-empty and compact.

Theorem. Let (X, π) be an sds. Let $x \in X$. Let $x \in \Lambda(x)$. Then $L(x) = J(x)$.

Proof. For simplicity of notation, let X be a metric space, and d a meter on it. Since $x \in \Lambda(x)$, there exists a sequence $\{t_n\}$ in R^+, $t_n \to \infty$ such that $xt_n \to x$. Let $y \in L(x)$. There exists a sequence $\{x_n\}$ in X, a sequence $\{s_n\}$ in R^+, $s_n \to \infty$ such that $x_n s_n \to y$ and $x_n \to xs$ for some $s \in R^+$. By making adjustments, we can suppose $s_n - t_n > n$ for every n. For every k fixed,

$$x_n(t_k - s) \to xs(t_k - s) = xt_k.$$

Let

$$d(x_n(t_k - s), xt_k) < 1/k \qquad \text{for every} \quad n \geq k.$$

Now,

$$d(x, x_n(t_n - s)) \leq d(x, xt_n) + d(xt_n, x_n(t_n - s))$$
$$< d(x, xt_n) + 1/n,$$

and so $x_n(t_n - s) \to x$. Finally

$$x_n(t_n - s)(s_n - t_n + s) = x_n s_n \to y,$$

and so $y \in J(x)$, etc.

In particular, if x is periodic, then $J(x) = L(x)$.

Remark. In the above theorem, if we weaken the hypothesis $x \in \Lambda(x)$ to $x \in J(x)$, then in general $L(x)$ and $J(x)$ are not the same.

Theorem. Let (X, π) be an sds. Let $x \in X$. Let $\omega \in \Lambda(x)$. Then

(a) $J(x) \subset J(\omega)$,
(b) $L(x) \subset J(\omega)$,
(c) $J(\omega) = D(\omega)$.

REFERENCES

[1] J. Auslander and P. Seibert, Prolongations and stability in dynamical systems, *Ann. Inst. Fourier, Grenoble* **14** (1964), 237–268.
[2] P. N. Bajaj, Start points in semidynamical systems, *Funkcial. Ekvac.* **13** (1971), 171–177.
[3] P. N. Bajaj, Connectedness properties of start points in semidynamical systems, *Funkcial. Ekvac.* **14** (1971), 171–175.

[4] P. N. Bajaj, Singular points in products of semidynamical systems, *SIAM J. Appl. Math.* **18** (1970), 282–286.

[5] N. P. Bhatia and G. P. Szegö, "Dynamical Systems, Stability Theory and Applications." Springer-Verlag, Berlin and New York, 1967.

[6] O. Hajek, Ordinary and asymptotic stability of noncompact sets, *J. Differential Equations* **11** (1972), 49–65.

[7] O. Hajek, Absolute stability of noncompact sets, *J. Differential Equations* **9** (1971), 496–508.

[8] J. R. Isbell, Uniform Spaces, Math. Surveys No. 12. Amer. Math. Soc., 1964.

[9] V. V. Nemytskii and V. V. Stepanov, "Qualitative Theory of Differential Equations." Princeton Univ. Press, Princeton, New Jersey, 1960.

[10] S. Willard, "General Topology." Addison-Wesley, Reading, Massachusetts, 1970.

[11] T. Ura, On the flow outside a closed invariant set; stability, relative stability and saddle sets, *Cont. Differential Equations* **3** (1964), 249–294.

When Do Lyapunov Functions Exist on Invariant Neighborhoods?

NAM P. BHATIA and M. NISHIHAMA
Division of Mathematics and Physics
University of Maryland, Baltimore, Maryland

The theory of characterization via Lyapunov functions of compact asymptotically stable sets of dynamical systems defined on locally compact metric spaces is essentially complete [1]. Krasovskii [2] extended such characterization of asymptotically stable critical points of differential systems to include the more general situation where a critical point has a neighborhood containing no other complete trajectories [2, Chap. 1]. Further notable developments in these directions have been obtained, for example, in [3] for autonomous differential systems and in [4] for dynamical systems on locally compact spaces. The basic characterization of isolated compact sets of dynamical systems via Lyapunov functions is contained in Theorem 1.3 (see also [5], where it was reported first). However, in contrast to the situation for asymptotically stable sets the characterizing Lyapunov functions are generally not available on any invariant neighborhoods of an isolated compact invariant set. (See [5] for an example.) In this direction, Theorem 1.4 was reported in [5]. Here we have eliminated the condition of invariance on the compact isolated set to obtain a generalization of Theorem 1.4. Our results even include results on the existence of tubes and ∞-tubes as neighborhoods of noncritical points that form the basis of the theory of parallelizable dynamical systems [1, Chap. IV]. We believe that our results bring the characterization of isolated compact sets via Lyapunov functions to a certain completion and can form the basis of classification of such sets.

1. Notation, Definition, and Some Known Results

X denotes a locally compact metric space and R the set of real numbers. A dynamical system or continuous flow on X, namely, a continuous function π from $X \times R$ into X satisfying $x0 = x$ and $(xt)s = x(t + s)$ for

197

all $x \in X$ and t, $s \in R$, is assumed given. (See [1] for this and related notation.) A set M in X is called invariant if $M = MR = \{xt : x \in M$ and $t \in R\}$.

The following definition is indeed Definition 2.2 of [5].

Definition 1.1. A compact set M in X is said to be isolated if there is a neighborhood U of M such that for every $x \in U - M$ at least one of the following holds:

(1.1.1)   There is a $t > 0$ such that $xt \notin \bar{U}$ and $x[0, t] \cap M = \varnothing$;

(1.1.2)   There is a $t < 0$ such that $xt \notin \bar{U}$ and $x[0, t] \cap M = \varnothing$.

For compact invariant sets the above definition is equivalent to the one given in [4], namely, that a compact invariant set is isolated if it is the largest invariant set in one of its neighborhoods. To show the generality of the above definition note the following.

Lemma 1.2. A noncritical point $x \in X$ is isolated. (This is the "only if" part of Lemma 2.3 in [5]; the "if" part does not hold.)

The following results were reported in [5].

Theorem 1.3. A compact set is isolated if and only if there is a neighborhood U of M and a continuous function φ from U into R satisfying

(i) $\varphi^{-1}(0) = \{x \in U : \varphi(x) = 0\} \supset M$,

(ii) $\varphi(xt) < \varphi(x)$ if $t > 0$ and $x[0, t] \subset (U - M)$.

For compact invariant sets we had moreover reported the following theorem [5, Theorem 3.8].

Theorem 1.4. A compact invariant set M possesses an invariant neighborhood U of M and a continuous function φ from U into R satisfying conditions (i) and (ii) of Theorem 1.2 if and only if M is isolated and there is a neighborhood V of M such that

$$(V - M) \cap J^+(M) \cap J^-(M) = \varnothing.$$

(Here for any $x \in X$, $J^+(x)$ and $J^-(x)$ denote the positive and negative prolongational limit set of x, respectively [1].)

2. Remarks

2.1. If M is a singleton $\{x\}$ and x is not critical, then by Lemma 1.2 x is isolated. The "only if" part of Theorem 1.3 is then essentially the Bebutov result on the existence of a tube containing x [1, Theorem 2.9].

2.2. It is known that for wandering points x, i.e., if $x \notin J^+(x)$, there exists an ∞-tube [1, Theorem 2.12]. However, this result is not contained in Theorem 1.4 as M is invariant and for a noncritical point x, the singleton $\{x\}$ is not invariant. At first sight it may appear that Theorem 1.4 cannot be strengthened beyond requiring that M is either compact invariant or a singleton, as is shown by the following example.

Example 2.3. Consider a dynamical system on $X = R$ containing exactly two critical points, say -1 and $+1$. Let $M = \{x, y\}$, where $-1 < x < y < +1$. One can easily see that M is a compact isolated set and that for any neighborhood V of M that does not contain the critical points the condition $(V - M) \cap J^+(M) \cap J^-(M) = \varnothing$ holds. However, as may easily be seen a continuous function φ satisfying conditions (i) and (ii) of Theorem 1.3 does not exist on any invariant neighborhood.

2.4. One can even give examples of connected compact sets M (necessarily not invariant) that are isolated and for which the condition of the "if part" of Theorem 1.4 holds, but a continuous function on an invariant neighborhood U of M with the requisite properties does not exist.

2.5. The proofs of Theorems 1.3 and 1.4 may be constructed by using methods used in, say, [4] and [6].

3. New Results

Definition 3.1. A set M in X is said to be t-convex (short for trajectory convex) if for any $x \in X$ and $t_1, t_2 \in R$, $t_1 < t_2$, the trajectory segment $x[t_1, t_2] \subset M$, whenever its endpoints xt_1 and xt_2 are in M.

One can easily verify that an invariant set is t-convex. In fact, one has the following characterization:

Lemma 3.2. A set M in X is t-convex if and only if $M = MR^+ \cap MR^-$. Here $MR^+ = \{xt: x \in M \text{ and } t \geq 0\}$ and $MR^- = \{xt: x \in M \text{ and } t \leq 0\}$.

Proof. The inclusion $M \subset (MR^+ \cap MR^-)$ holds for arbitrary sets M in X. First let M be t-convex. Let $x \in (MR^+ \cap MR^-)$. There is then a $t_1 \leq 0$ and $t_2 \geq 0$ such that $xt_1 \in M$ and $xt_2 \in M$. By t-convexity, we have $x[t_1, t_2] \subset M$. Since $x = x0 \in x[t_1, t_2]$, we conclude $x \in M$. This proves the "only if" part. Now let $M = MR^+ \cap MR^-$. If $x \in X$ and $t_1 < t_2$ in R are such that $xt_1 \in M$ and $xt_2 \in M$, then indeed $x[t_1, t_2] \subset (MR^+ \cap MR^-)$. Hence $x[t_1, t_2] \subset M$ and M is t-convex. This concludes the proof.

The interior of a t-convex set is t-convex. It is not true that the closure of an arbitrary or even an open t-convex set is t-convex. However, an arbitrary positively invariant set is t-convex and so are its closure, interior, boundary, and complement. One can also see that a singleton $\{x\}$ consisting of a noncritical nonperiodic point x is t-convex and hence also a wandering point is t-convex.

We report now the following generalization of Theorem 1.4.

Theorem 3.3. A compact set M in X has an invariant neighborhood U and a continuous function φ from U into R satisfying conditions (i) and (ii) of Theorem 1.2 if and only if M is t-convex, isolated, and has a neighborhood V such that $(V - M) \cap D^+(M) \cap D^-(M) = \varnothing$.

Remark 3.4. It would appear from Theorem 3.3 that the sections of an ∞-tube containing a wandering point are t-convex.

Theorem 3.3 can be proved by the same technique as Theorem 1.4. The only new element is the assertion in the "only if" part that M is t-convex. To see this, assume that there is an invariant neighborhood U of M with a continuous function φ on U satisfying conditions (i) and (ii) of Theorem 1.3. Let $x \in X$ and $t_1 < t_2$ in R be such that xt_1 and xt_2 are in M. Then by the invariance of U, we have $x[t_1, t_2] \subset U$. Conditions (i) and (ii) on φ now imply that $\varphi(xt) = 0$ for $t_1 \leq t \leq t_2$. If for some $t \in (t_1, t_2)$ one had $xt \notin M$, then there would exist an $\varepsilon > 0$ such that $t_1 < t < t + \varepsilon < t_2$ and $x[t, t + \varepsilon] \subset U - M$ (this follows from compactness of M). But then condition (ii) implies $\varphi(x(t + \varepsilon)) < \varphi(xt)$. This contradicts the earlier conclusion that $\varphi(xt) = 0$ for $t_1 \leq t \leq t_2$. Hence $x[t_1, t_2] \subset M$ and M is t-convex.

ACKNOWLEDGMENTS

The first author acknowledges partial support of the National Science Foundation through Grant GP-39057. The second author wishes to thank Dr. Morton Baratz, Vice-Chancellor, UMBC, for support of his research.

REFERENCES

[1] N. P. Bhatia and G. P. Szegö, "Stability Theory of Dynamical Systems" (Grundleheren d. Math. Wiss. No. 161), Springer-Verlag, Berlin and New York, 1970.

[2] N. N. Krasovskii, "Stability of Motion." Stanford Univ. Press, Stanford, California, 1963.

[3] F. Wesley Wilson, Jr. and J. A. Yorke, Lyapunov functions and isolating blocks, *J. Differential Equations* 13 (1973), 106–123.

[4] R. C. Churchill, Isolated invariant sets in compact metric spaces, *J. Differential Equations* 12 (1972), 330–352.

[5] N. P. Bhatia, Dynamical flow near a compact invariant set in "Équations differentelles et fonctionnelles non linéaires" (P. Janssens, J. Mawhin, and N. Rouche, eds.), pp. 519–312. Hermann, Paris, 1973.

[6] N. P. Bhatia, On asymptotic stability in dynamical systems, *Math. Syst. Theory* 1 (1967), 113–128.

The "Simplest" Dynamical System*

TIEN-YIEN LI
Department of Mathematics
The University of Utah, Salt Lake City, Utah

JAMES A. YORKE
Institute for Fluid Dynamics and Applied Mathematics
University of Maryland, College Park, Maryland

1. Introduction

Perhaps the simplest "dynamical process" is obtained by iterating a function $F: J \to J$, where F is a continuous function and J an interval. When we iteratively define $F^{n+1}(x)$ as $F(F^n(x))$, we ask questions about what we can expect of the long-term or asymptotic behavior of $\{F^n(x_0)\}$ for $x_0 \in J$. These questions are particularly interesting because of the rich and surprisingly complex structure that is revealed in the investigation of these questions. Our purpose here is to describe some of the general results we have proved and more importantly to describe our observations (results of simple computer studies) concerning the simplest case

$$F_a(x) = ax(1 - x), \qquad X \in J = [0, 1],$$

for various values of $a \in [0, 4]$. (These are the values of a for which $F_a: J \to J$.) Our purpose in studying how we may expect the sequence $\{F^n(x)\}$ to behave is to get a new glimmer of understanding of the possible analytic regularities in the chaotic, turbulent, unstable, irregular processes of such greater complexity that surround us.

2. Applications

Our initial investigations were stimulated by a fascinating series of Lorenz [1-4], who found that certain turbulent (or irregular) fluid flows could be reduced, through a series of careful approximations, to studies of the scalar problem of studying sequences $\{F^n(x)\}$ as we do here. In his

* This research was partially supported by the National Science Foundation under grant GP 31886X

case $F^n(x)$ can be thought of as being the magnitude of the nth wave to pulse through the system. He found that these sequences often behaved irregularly in much the same manner as had been reported for the original experimental situations. We feel that studies of these sequences can give significant insight into the nature of turbulence.

Sequences of this type are also quite important in the study of mathematical ecology. For example, recent investigations of Oster *et al.* have shown that discrete-time models can give excellent descriptions of the interactions of insect populations in laboratory environments. To understand the behavior of *systems* it is necessary to understand the dynamics of a single population, which we show can at times be expected to be very complicated.

3. Predicting Chaos

The following theorem (which we prove elsewhere [5]) shows that if certain simple qualitative hypotheses are satisfied, then the asymptotic description of sequences $\{F^n(x)\}$ is necessarily complicated. First we state the result in terms of a population. Suppose that under certain laboratory situations, the size (denoted x_n) of the nth generation is a continuous function of the previous generation size, that is,

$$x_n = F(x_{n-1}).$$

Suppose further that when we start with a first generation of size x_1 for some $x_1 > 0$ we find the population grows for two successive generations, that is, $x_1 < x_2 < x_3$, but then the population drops to a level lower than the initial level, that is, $x_4 \leq x_1$. With no additional hypotheses the following result shows that there must be an uncountable set S of initial population levels with "irregular" behavior. For example, for $y \in S$, the sequence $F^n(y)$ is not periodic in n and does not tend asymptotically to a periodic point. Here we call $x \in J$ a *periodic point of period* k if $F^k(x) = x$ and $F^m(x) \neq x$ for all $1 \leq m < k$. We say $x \in J$ is *asymptotically periodic* if there is a periodic point p for which

$$F^n(x) - F^n(p) \to 0 \qquad \text{as} \quad n \to \infty.$$

Theorem. Let J be an interval and let $F: J \to J$ be continuous. Assume there is a point $a \in J$ for which the points $b = F(a)$, $c = F^2(a)$, and $d = F^3(a)$ satisfy

$$d \leq a < b < c \qquad (\text{or} \quad d \geq a > b > c).$$

Then

(1)   For every $k = 1, 2, \ldots$ there is a period point in J having period k.

Furthermore,

(2)   There is an uncountable set $S \subset J$ (containing no periodic points), which satisfies the following conditions:

   (A)   For every $p, q \in S$ with $p \neq q$,

$$\limsup_{n \to \infty} |F^n(p) - F^n(q)| > 0, \qquad \liminf_{n \to \infty} |F^n(p) - F^n(q)| = 0.$$

   (B)   If $y \in S$ then y is not asymptotically periodic.

Remark.   Notice that if there is a periodic point with period 3, then the hypothesis of the theorem will be satisfied.

4. Computation Observation

Consider the simplest nonlinear function

$$F_a(x) = ax(1 - x), \qquad x \in J = [0, 1],$$

for $a \in [0, 4]$. The asymptotic behavior of F_a is not well understood but it can be shown that F_a has asymptotically stable periodic points for a large set of a (see, for example, Li and Yorke [5] for a discussion of this stability condition). For instance, there is a set of values $a_n \uparrow 4$ for which $\frac{1}{2}$ is an asymptotically stable point of period n with $F_{a_2}^1(\frac{1}{2}) > \frac{1}{2}$ and

$$F_{a_n}^2(\tfrac{1}{2}) < F_{a_n}^3(\tfrac{1}{2}) < \cdots < F_{a_n}^n(\tfrac{1}{2}) = \tfrac{1}{2}.$$

It can be seen that whenever $\frac{1}{2}$ is a periodic point of F_a, it is an asymptotically stable periodic point since $(d/dx)F_a(x) = 0$ for $x = \frac{1}{2}$ and so $(d/dx)F_a{}''(x) = 0$ for $x = \frac{1}{2}$. From continuous dependence of $F_a{}^k(\frac{1}{2})$ on a, it can be shown for $n > 2$ that

$$a_n = \inf\{a > 3 : F_a{}^k(\tfrac{1}{2}) < \tfrac{1}{2} \text{ for each } k = 2, \ldots, n\}.$$

Let L_a denote the attracting limit set for $F_a(x) = ax(1 - x)$. Professor Robert May pointed out to us that L_a consists of a periodic orbit of period 2^n (so L_a has 2^n points) for low values of a and that n increases to ∞ as a approaches some critical value a_∞. Through an intensive computation we constructed Fig. 1 to obtain a more graphic view of this behavior. This figure represents our findings based primarily on computer experiment rather than rigorous proof.

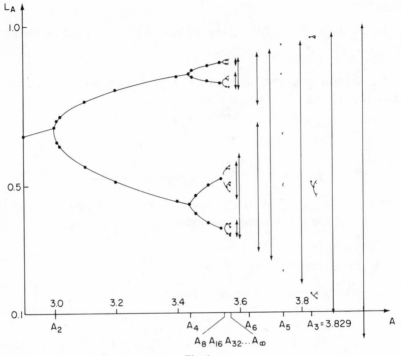

Fig. 1

REFERENCES

[1] E. N. Lorenz, The problem of deducing the climate from the governing equations, *Tellus* **16** (1964), 1–11.
[2] E. N. Lorenz, Deterministic nonperiodic flows, *J. Atmos. Sci.* **20** (1963), 130–141.
[3] E. N. Lorenz, The mechanics of vacillation, *J. Atmos. Sci.* **20** (1963), 448–464.
[4] E. N. Lorenz, The predictability of hydrodynamic flow, *Trans. N.Y. Acad. Sci. Ser. II* **25** (1963), 409–432.
[5] T. Y. Li and J. Yorke, Period three simplest chaos, *Amer. Math. Monthly* (to appear).

Continuous Operators That Generate Many Flows

COKE S. REED

Mathematics Department
Auburn University, Auburn, Alabama

Suppose that G is a continuous transformation from E^2 into E^2. T is a dynamical system that solves G means that T is a continuous transformation from $E^1 \times E^2$ onto E^2 such that if each of x and y is a number and p is a point of the plane, then

$$T(x, T(y, p)) = T(x + y, p) \quad \text{and} \quad \dot{T}(p) = G(p).$$

Anatol Beck has given an example of a continuous transformation G from the plane into itself such that there exist uncountably many dynamical systems on the plane that solve G. The purpose of this paper is to define a continuous transformation G from the plane into itself with the property that if p and q are two points of the plane, then there is a dynamical system T with the property that there is a trajectory ψ of T such that both p and q are in $\text{cl}(\psi)$ and T solves G.

Let S denote the square with vertices $(0, 0)$, $(0, 1)$, $(1, 0)$, and $(1, 1)$ and let I denote the interior of S. Suppose that α is a point of S on either the bottom or left side, β a point of S on either the top or right side, and γ a point of I above and to the right of α and below and to the left of β. Suppose that for each $w \in (0, 1)$, t_w is a positive number and F_w a continuous transformation from $[0, t_w]$ into $I \cup S$. F denotes the collection of all F_w. Now suppose that G is a continuous transformation from $I \cup S$ into E^2 such that for each point p of S, $G(p) = 0$. F is said to be an $[\alpha, \beta, \gamma, G]$ collection if and only if F satisfies the following seven conditions:

(1)   For each $w \in (0, 1)$, $F_w[0, t_w]$ is a strictly increasing continuous graph;

(2)   For each $w \in (0, 1)$, $F_w(0)$ is a point of S on the same side of S as α, $\|F_w(0)\| = 0$, and $F_w(t_w)$ is a point of S on the same side of S as β;

(3)   There is a $w \in (0, 1)$ and a $t \in (0, t_w)$ such that $F_w(0) = \alpha$, $F_w(t_w) = \beta$, and $F_w(t) = \gamma$;

(4)   If r is a point of I or a point on the α or β side of S, then there is exactly one pair (w, t) such that $F_w(t) = r$;

(5) If $w \in (0, 1)$, $t \in (0, t_w)$, and $\varepsilon > 0$, then there is a $\delta > 0$ such that if $|w' - w| < \delta$ and $|t' - t| < \delta$, then $t' \in (0, t_{w'})$ and $\|F_w(t) - F_{w'}(t')\| < \varepsilon$;

(6) If r is a point of S not on the same side as α or β, $J > 0$, and $\varepsilon > 0$, then there is a $\delta > 0$ such that if $F_w(t)$ is a point of I and $\|F_w(t) - r\| < \delta$ then $0 < t - J < t + J < t_w$ and $\|F_w(t') - r\| < \varepsilon$ for each $t' \in (t - J, t + J)$; and

(7) For each point $F_w(t)$ of I, $F_{w'}(t) = G(F_w(t))$.

Suppose that G is a transformation from $I \cup S$ into E^2. G is said to be acceptable if and only if for each point (x, y) of I, there exist four numbers A, B, C, and D such that

$$\max[0, x - \tfrac{1}{10}] < A < x, \qquad \max[y - \tfrac{1}{10}] < B < y,$$

$$x < C < \min[x + \tfrac{1}{10}, 1], \qquad y < D < \min[y + \tfrac{1}{10}, 1],$$

and such that if

$$\alpha \in [(0, 0), (A, 0)] \cup [(0, 0), (0, B)], \qquad \beta \in [(C, 1), (1, 1)] \cup [(1, D), (1, 1)],$$

then there exists an $[\alpha, \beta, (x, y), G]$ collection F.

In order to define an acceptable G, it will be necessary to establish certain notation. Let K denote the "middle $\tfrac{1}{20}$" Cantor set in $[0, 1]$. Let g denote a function on $[0, 1]$ such that:

(1) g''' exists;
(2) $g(x) = 0$ for each $x \in K$;
(3) $g(x) > 0$ for each $x \in [0, 1]\backslash K$;
(4) For each nondegenerate component of $[0, 1]\backslash K$, $\int_a^b g < \tfrac{1}{20}$; and
(5) $\int_0^1 g = 1$.

Set $f(x) = \int_0^x g$. This f is of the type first introduced by Beck. For each $(x, y) \in I$ and each number t such that

$$[f(f^{-1}(x) + t), f(f^{-1}(y) + t)] \in I \cup S,$$

set

$$\varphi(t, (x, y)) = (f(f^{-1}(x) + t), f(f^{-1}(y) + t)).$$

For each $p \in I$, let $t(p) \geq 0$ and $\tau(p) \leq 0$ be such that $\varphi(t, p)$ and $\varphi(\tau, p)$ are on S. Set $T(p) = \min(|t(p)|, |\tau(p)|)$. For each point $p \in I \cup S$, set

$$G(p) = \begin{cases} \dot{\varphi}(p) & \text{if} \quad \|\dot{\varphi}(p)\| \leq 2(T(p))^{1/2}, \\ (2(T(p))^{1/2}/\|\dot{\varphi}(p)\|)\dot{\varphi}(p) & \text{if} \quad 2(T(p))^{1/2} < \|\dot{\varphi}(p)\|. \end{cases}$$

It can be shown that G is acceptable. It is, however, beyond the scope of this work to establish this fact. Define G_1 and G_2 to be the transformations from E^2 into E^1 such that for each $p \in I \cup S$, $G(p) = (G_1(p), G_2(p))$.

G will now be extended to the entire plane as follows: If each of J and K is an integer and $(x, y) \in I$, then set

$$G(x + 2J, y + 2K) = G(x, y),$$

$$G(x + 2J + 1, y + 2K) = (G_2(1 - y, x), -G_1(1 - y, x)),$$

$$G(x + 2J, y + 2K + 1) = (-G_2(y, 1 - x), G_1(y, 1 - x)),$$

$$G(x + 2J + 1, y + 2K + 1) = (-G_1(1 - x, 1 - y), -G_2(1 - x, 1 - y)).$$

One can think of G as a vector field induced on the plane by covering the plane with squares of area one, each of which contains a picture of a translation and rotation of the vector field induced on $I \cup S$ by G.

Suppose now that p and q are two points of the plane. There is a dynamical system T and a trajectory ψ of T such that both p and q are in $\mathrm{cl}(\psi)$. An outline of the construction of such a T will now be given. Assume, without loss of generality, that p is in I or p is on the lower side of S.

Let a, b, and c denote three points of the plane such that if l_1 is the line interval with endpoints a and b and l_2 is the line interval with endpoints b and c, then the following five conditions are satisfied:

(1)   Each of a, b, and c has integral coordinates;
(2) l_1 is horizontal and l_2 is vertical;
(3) $\|a - p\| > 5$, $\|a - q\| > 5$, $\|b - p\| > 5$, and $\|b - q\| > 5$;
(4)   The distance from p to $l_1 \cup l_2$ is less than one and the distance from q to $l_1 \cup l_2$ is less than one; and
(5)   The abcissa of b is odd.

Let W denote the collection of all squares in the plane to which s belongs if and only if the area of s is one and each corner of s has integral coordinates. Let w_1 denote the subcollection of W to which the square s belongs only in case there is a point of $l_1 \cup l_2$ on s. For each integer $n > 1$, inductively define w_n to be the subcollection of $W \backslash (w_1 \cup w_2 \cup \cdots \cup w_{n-1})$ to which s belongs only in case there is a point of a member of w_{n-1} on s. For each integer n, let V_n denote the set of all points r such that r is on a member of V_n or r is in the interior of a member of V_n. Let Y_n denote the interior of V_n and let Z_n denote the boundary of V_n.

Notice that there is a dynamical system T such that:

(1) If neither p nor q is in $l_1 \cup l_2$, then q is on the positive semi-trajectory of p;

(2) If p is not in $l_1 \cup l_2$ and q is in $l_1 \cup l_2$, then there is a trajectory of T that contains p and has $l_1 \cup l_2$ as its Ω-limiting set;

(3) If p and q are both in $l_1 \cup l_2$, then there is a trajectory of T that has $l_1 \cup l_2$ as its Ω-limiting set;

(4) Each point of $(l_1 \cup l_2) \cup (\bigcup_{n=1}^{\infty} Z_n)$ is a rest point of T;

(5) Either, for each point r of Y_1, $l_1 \cup l_2$ is the α-limiting set of r and Z_1 is the Ω-limiting set of r, or vice versa;

(6) Either, for each point r of Y_n $(n > 1)$, Z_{n-1} is the α-limiting set of r and Z_n is the Ω-limiting set of r, or vice versa.

Consider the family of theorems of the following form: There exists a continuous transformation G from the space S into S such that if p and q are points of S, then there is a dynamical system T such that T solves G and T has a trajectory ψ such that both p and q are on $cl(\psi)$. The following questions are open:

(1) In case $S = E^2$, can "$cl(\psi)$" be replaced by "ψ" (the author once incorrectly claimed that he could solve this problem)?

(2) In case the answer to (1) is false, can "$cl(\psi)$" be replaced by "ψ" for some other space S?

(3) Is there an S such that the phrase "with no rest point" can be inserted after the phrase "dynamical system T"?

Existence and Continuity of Liapunov Functions in General Systems

PABLO M. SALZBERG*

Instituto de Matemática
Universidad Católica de Chile, Santiago, Chile

Introduction

Liapunov's original "second method" has been varied in several ways, for instance, (a) with respect to regularity conditions imposed on the Liapunov function, like continuity, differentiability, etc.; (b) with respect to the monotonicity condition, which can be replaced by weaker ones; (c) with respect to the properties of the system under investigation, such as stability of various types, boundedness, etc. In addition, there has been a proliferation of ramifications of the original theory and very often we have so many requirements on the functions that we lose completely the notion of the exact role of each of the conditions. So, it is time to analyze what is the underlying idea of the notion of a Liapunov function, and it seems to us that a unified theory should: (1) *involve a minimal number of objects,* which should be simple and related among themselves in a simple way, and (2) *cover a maximal number of stability or other notions.*

The first step in this direction is due to Bushaw [1]. Later, inspired by Bushaw's paper, two other more refined theories appeared simultaneously, namely those by Pelczar [2] and Seibert [3]. In what follows, we shall refer to Seibert's version because it is the closest to requirements (1) and (2).

The context is that of a set X (the phase space) endowed with a preorder ϕ (the flow),** and two nonempty collections of sets $\mathscr{D}$, $\mathscr{E} \subset 2^X - \{0\}$ (which generalizes the notion of neighborhood filters).

Summarizing, we call $\Sigma = (X, \phi, \mathscr{D}, \mathscr{E})$ *a system.*

Now we say the quasifilter $\mathscr{E}$ is coarser than $\mathscr{D}$ and write $\mathscr{E} \prec \mathscr{D}$ if $\forall E \in \mathscr{E}$, $\exists D \in \mathscr{D}$ such that $D \subset E$. If $\mathscr{E} \prec \mathscr{D}$ and $\mathscr{D} \prec \mathscr{E}$ we say that $\mathscr{E}$ and $\mathscr{D}$

---

* Present address: Departmento de Matemáticas, Universidad Simón Bolivar, Caracas, Venezuela.

** We use the symbol ϕ both for the preorder and for the mapping from X into 2^X defined by $x\phi y$ iff $y \in \phi(x)$.

are equivalent and denote this fact by $\mathscr{D} \rightarrowtail \mathscr{E}$. In this general framework we introduce stability as follows:

Definition. Let there be given a system $\Sigma = (X, \phi, \mathscr{D}, \mathscr{E})$. We say that Σ is stable if $\mathscr{E} \prec \phi(\mathscr{D})$.

To explain within this setting what is meant by a Liapunov function, let $v: X \to [0, \infty]$ and define for each $r > 0$, the set $S_v^r = \{x \in X \,|\, v(x) < r\}$, and for every function v, the collection of sets $\mathscr{S}_v = \{S_v^r \,|\, S_v^r \neq 0\}$, which is a quasifilter unless v is identically ∞.

Definition. The function v is called a *Liapunov function* (or briefly an *L-function*) for a given system $\Sigma = (X, \phi, \mathscr{D}, \mathscr{E})$ if the following conditions are satisfied:

(L.1) $$\mathscr{E} \prec \mathscr{S}_v,$$

(L.2) $$\mathscr{S}_v \prec \mathscr{D},$$

(L.3) $$\phi(S_v^r) \subset S_v^r \qquad (r > 0).$$

1. On the Existence of Liapunov Functions

The first main result of the theory was presented in a joint paper by the author and Seibert [4] and can be formulated as follows:

Theorem 1. The following condition is necessary and sufficient for the existence of an *L*-function for $(X, \phi, \mathscr{D}, \mathscr{E})$: There exists a quasifilter $\mathscr{F}$ satisfying the conditions:

(I) $$\mathscr{E} \prec \mathscr{F} \prec \mathscr{D}.$$

(II) $\mathscr{F}$ is equivalent to a nested and countable quasifilter (admissibility condition), and at least one of the following:

(IIIa) $(X, \phi, \mathscr{D}, \mathscr{F})$ is stable,

(IIIb) $(X, \phi, \mathscr{F}, \mathscr{E})$ is stable,

(IIIc) $(X, \phi, \mathscr{F}, \mathscr{F})$ is stable.

Moreover, if the function in question exists, one can always find a quasifilter $\mathscr{F}$ satisfying (I), (II), and (IIIc).

The reader is referred to the papers cited above for details and examples.

Although the admissibility condition is essential here (to handle only one L-function), it can be relaxed. The references in this direction are the papers by Auslander [5] and by the author and Seibert [6].

2. Semicontinuity and Continuity of Liapunov Functions

In the preceding section, we defined for each function $v: X \to R^{\#*}$ an admissible quasifilter $\mathscr{S}_v$, and realized how Liapunov-type properties of v can be naturally phrased in terms of $\mathscr{S}_v$. Conversely, given an admissible quasifilter $\mathscr{F}$ indexed by some function, say $F: R^{\#} \to \mathscr{F}$, two dual functions were introduced in [3, Sections 4, 5], one of which, $v_F: X \to R^{\#}$, was defined by

(2.1) $\qquad v_F(X) := \begin{cases} \infty, & \text{if } \{r \mid \phi(x) \subset F(r)\} \text{ is empty,} \\ \inf\{r \mid \phi(x) \subset F(r)\}, & \text{otherwise.} \end{cases}$

Again, Liapunov-type properties of v_F are implied by suitable conditions on $\mathscr{F}$. We have thus established a correspondence between functions and quasifilters that is the core of the unified theory developed in the papers cited above.

Our next concern shall be to analyze closely the transference of properties among the elements involved in this correspondence when X is endowed with a topology. In particular, we will look for those properties on the quasifilter that yield (semi-) continuity of the associated L-functions.

Unfortunately, since (2.1) furnishes the bridge from quasifilters to L-functions, the required properties will depend on the order structure of the nested countable quasifilter equivalent to $\mathscr{F}$ that we choose, rather than on the quasifilter $\mathscr{F}$ itself. Therefore, we need a suitable notation for nested countable quasifilters that also reflects its order structure. This will be provided by the following

Lemma. A quasifilter $\mathscr{F}$ is admissible if and only if there exists a function $F: I \subset R^{\#} \to \mathscr{F}$ satisfying the following conditions:

(2.2) $\qquad\qquad\qquad\qquad \bar{I} = R^{\#},$

(2.3) $\qquad\qquad\qquad\qquad F(I) \rightarrowtail \mathscr{F},$

(2.4) $\qquad\quad \forall r, r' \in I, \quad r < r' \quad \text{implies} \quad F(r) \subset F(r').$

---

* $R^{\#} = [0, \infty]$.

The function F appearing in the lemma is called a *base* of $\mathscr{F}$.

Now, given an admissible quasifilter $\mathscr{F}$, we say a base F of $\mathscr{F}$ is *invariant under* ϕ if

(2.5) $\phi \circ F(r) \subset F(r), \qquad \forall r \in I.$

Let X be endowed with a topology (we assume all topological spaces are Hausdorff).

Definition. Given an admissible quasifilter $\mathscr{F}$, we call a base F of $\mathscr{F}$ *upper normal* (*lower normal*), if condition (2.6) [resp. (2.7)] holds:

(2.6) $\forall r, r' \in I, \quad r < r' \Rightarrow F(r) \subset \text{int}(F(r')),$

(2.7) $\forall r, r' \in I, \quad r < r' \Rightarrow \overline{F(r)} \subset F(r')$

(where $\overline{F(r)}$ denotes the closure of $F(r)$).

Furthermore, F is *normal* if both (2.6) and (2.7) hold.

To illustrate the notions just introduced, we shall exhibit some elementary but clarifying examples.

Example 1. Let $v \colon X \to R^{\#}$ be any function. A natural base function of $\mathscr{S}_v$ is given by $S(r) = S_v{}^r$. Hence, we shall refer to $\mathscr{S}_v$ as being invariant or (upper, lower) normal when S has the property in question.

Example 2. Set $X = R_2$ and let $\mathscr{F} = \mathscr{N}_0$ be the neighborhood filter of the origin. Then the base F of $\mathscr{F}$ defined by $F(r) = \{(x, y) \mid |x| \le r, |y| \le r\}$ is normal.

We may now formulate the main results of this paper.

Theorem 2. There exists an upper semicontinuous L-function for $(X, \phi, \mathscr{D}, \mathscr{E})$ iff it is possible to find a quasifilter $\mathscr{F}$ that simultaneously satisfies

(2.8) $\mathscr{E} \prec \mathscr{F} \prec \mathscr{D},$

and

(2.9) $\mathscr{F}$ admits an invariant and upper normal base.

In duality to theorem 2, we claim:

Theorem 3. There exists a lower semicontinuous L-function for $(X, \phi,$ $\mathscr{D}, \mathscr{E})$ iff it is possible to find a quasifilter $\mathscr{F}$ that simultaneously satisfies

(2.10) $$\mathscr{E} \prec \mathscr{F} \prec \mathscr{D},$$

and

(2.11) $\mathscr{F}$ admits an invariant and lower normal base.

Theorem 4. There exists a continuous L-function for $(X, \phi, \mathscr{D}, \mathscr{E})$ iff there exists a quasifilter $\mathscr{F}$ admitting an invariant and normal base, that moreover satisfies $\mathscr{E} \prec \mathscr{F} \prec \mathscr{D}$.

Example 3. Consider the plane autonomous system

$$\dot{x} = 0, \qquad \dot{y} = y(x - y)(x + y)$$

(Fig. 1). Here (R_2, ϕ) is $\mathscr{N}_0$ stable. A normal and invariant base of $\mathscr{N}_0$ is given in Example 2. Hence, there exists a continuous L-function with respect to $(R_2, \phi, \mathscr{N}_0)$.

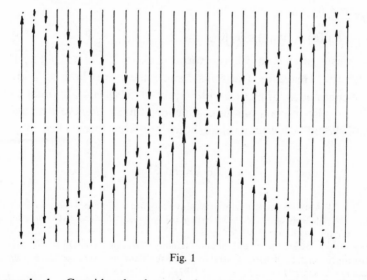

Fig. 1

Example 4. Consider the dynamical system on the planar half strip

$$X = \{(x, y) \in R_2 \,|\, x \geq 0, \, |y| \leq 1\}$$

as suggested by Fig. 2, where dots denote critical points and arrows indicate the direction of the flow along the regular orbits. This flow (ϕ) furnishes an example of stability in the sense of Liapunov for which it is

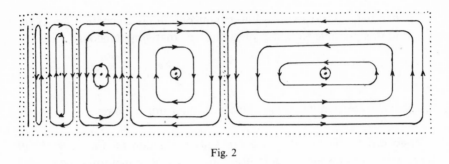

Fig. 2

possible to find both upper and lower semicontinuous L-functions, but there exists no continuous one. Indeed, let $Y = \{(0, y) \mid |y| \leq 1\}$ [which is stable in the sense of Liapunov, i.e., $(R_2, \phi, \mathcal{N}_Y)$ is stable]. Now, since the interior of each square (in Fig. 2) is open and invariant, one may easily find an invariant upper-normal base of $\mathcal{N}_Y$. To this effect, take for instance

$$F(\beta) = \{(x, y) \in R_2 \mid 0 \leq x < \frac{1}{n}, |y| \leq 1\} \qquad \text{if} \quad \frac{1}{(n+1)} < \beta \leq \frac{1}{n},$$

for $n = 1, 2, \ldots$, and

$$F(\beta) = \{(x, y) \in R_2 \mid 0 \leq x < 1, |y| < 1\} \qquad \text{if} \quad \beta > 1.$$

[Analogously, an invariant lower-normal base of $\mathcal{N}_Y$ is defined by $F'(\beta) = \overline{F(\beta)}$, $\beta \in R^{\#}$.] Hence, by Theorems 2 and 3 there exist both an upper- and a lower-continuous L-function with respect to $(X, \phi, \mathcal{N}_Y)$. In order to prove that there exists no continuous L-function, simply observe that any such function should be constant on the boundary, which contradicts $\mathcal{N}_0 \prec \mathcal{S}_v$.

REFERENCES

[1] Bushaw, D., A stability criterion for general systems, *Math. Systems Theory* **1** (1967), 79–88.
[2] Pelczar, A., Stability of sets in pseudodynamical systems, *Bull. Acad. Polon. Sci., Ser. Math. Astron. Phys.* **19** (1971), 13–17, 951–957.
[3] Seibert, P., A unified theory of Liapunov stability, *Funkcial. Ekvac.* **15** (1972), 139–147.
[4] Salzberg, P., and P. Seibert, A necessary and sufficient condition for the existence of a Liapunov function, *Funkcial. Ekvac.* **16** (1973), 97–101.
[5] Auslander, J., Filter stability in dynamical systems, *Tech. rep.* **73–62**, Univ. of Maryland, 1973.
[6] Salzberg, P., and P. Seibert, Remarks on a universal criterion for Liapunov stability, *Funkcial. Ekvac.* **18** (1975), 1–4.

Chapter 6: ORDINARY DIFFERENTIAL AND VOLTERRA EQUATIONS

Stability under the Perturbation by a Class of Functions

JUNJI KATO and TARO YOSHIZAWA
Mathematical Institute
Tohoku University, Sendai, Japan

Among stabilities under the perturbation by a class of functions, total stability has been well studied. One of these stability properties is integral stability, and another is M-stability, that is, stability under the perturbation that is small in the sense of the norm $\|\cdot\|_M$ defined by

$$\|h\|_M = \sup_{t \geq 0} \int_t^{t+1} |h(s)| \, ds.$$

The purpose of this chapter is to discuss relationships among totally asymptotic stability (hereafter called TAS for brevity), integrally asymptotic stability (IAS), M-asymptotic stability (MAS), and uniformly asymptotic stability (UAS).

The relations MAS $\Rightarrow$ TAS, MAS $\Rightarrow$ IAS, TAS $\Rightarrow$ UAS, and IAS $\Rightarrow$ UAS will be clear from the definitions given later, where $\Rightarrow$ stands for "implies." However, the relation UAS $\Rightarrow$ TAS or

$$\text{UAS} \Rightarrow \text{TS} \tag{1}$$

has been unsolved for a long time, where TS means total stability, while it is well known by the aid of Liapunov's second method that (1) is true under the uniform Lipschitz condition. For almost-periodic systems, (1) is verified under the condition that for every system in the hull, solutions are unique for initial value problems [3]. We shall call this condition uniqueness in the hull. In [2], uniqueness in the hull has been weakened, and Kato has shown that (1) holds true if uniformly asymptotic stability is uniform in the hull in some sense (UASH; see Definition 6). Furthermore, Yoshizawa [4] and Kato [2] have shown that UAS $\Rightarrow$ UASH for periodic systems without any additional condition and for almost-periodic systems with uniqueness in the hull. We can also find an example in [2] that shows that (1) is not necessarily true even for an almost-periodic system. These results were obtained for functional differential equations.

On the other hand, for ordinary differential equations, recently Chow and Yorke [1] have constructed a Liapunov function that gives a necessary and sufficient condition for IAS and by using this Liapunov function, they have shown that

$$\text{IAS} \Rightarrow \text{MAS},\tag{2}$$

and hence IAS is equivalent to MAS. However, the method of construction of the Liapunov function employed in [1] is not effective for functional differential equations. Here we shall give a different proof for (2) that is also effective for functional differential equations. Now we shall consider a system

$$x' = f(t, x), \qquad x, f \in R^n,\tag{3}$$

where R^n denotes the n-dimensional vector space and $f(t, x)$ is continuous on $I \times S_c$, $I = [0, \infty)$ and $S_c = \{x; |x| < c\}$. Denoting by $C(I, R^n)$ the set of all continuous functions on I with values in R^n, let B be a Banach space $\subset C(I, R^n)$ with the norm $\|\cdot\|_B$. In the following, $\varphi(t)$ is a solution of (3) such that $\sup_{t \geq 0} |\varphi(t)| \leq c^* < c$, and $y(t, t_0, y_0)$ denotes a solution through (t_0, y_0) of the system

$$y' = f(t, y) + h(t), \qquad h \in B.\tag{4}$$

Definition 1. The solution $\varphi(t)$ of (3) is said to be stable under B perturbations (called BS), if for any $\varepsilon > 0$ there exists a $\delta(\varepsilon) > 0$ such that $t_0 \geq 0$, $|\varphi(t_0) - y_0| < \delta(\varepsilon)$, and $\|h\|_B < \delta(\varepsilon)$ imply $|\varphi(t) - y(t, t_0, y_0)| < \varepsilon$ for all $t \geq t_0$.

Let T, L, and M be the Banach spaces $\subset C(I, R^n)$ with the norms $\|\cdot\|_T$, $\|\cdot\|_L$, and $\|\cdot\|_M$, respectively, where

$$\|h\|_T = \sup_{t \geq 0} |h(t)|, \qquad \|h\|_L = \int_0^\infty |h(t)| \, dt.$$

Then stability under B perturbations with $B = \{0\}$, $B = T$, $B = L$, or $B = M$ corresponds to uniform stability (US), total stability (TS), integral stability (IS), or M-stability (MS).

Definition 2. The solution $\varphi(t)$ of (3) is said to be attracting under B perturbations, if there exists a $\delta_0 > 0$ and for any $\varepsilon > 0$ there exists a $\tau(\varepsilon) > 0$ and an $\eta(\varepsilon) > 0$ such that $t_0 \geq 0$, $|\varphi(t_0) - y_0| < \delta_0$, and $\|h\|_B < \eta(\varepsilon)$ imply $|\varphi(t) - y(t, t_0, y_0)| < \varepsilon$ for all $t \geq t_0 + \tau(\varepsilon)$.

Definition 3. The solution $\varphi(t)$ of (3) is said to be asymptotically stable under B perturbations (called BAS), if it is stable under B perturbations and is attracting under B perturbations. If $B = \{0\}$, $B = T$, $B = L$, or $B = M$, this gives the definition of UAS, TAS, IAS, or MAS, correspondingly.

Moreover, we shall give the following two definitions, which characterize asymptotic stability under B perturbations.

Definition 4. We say that $\varphi(t)$ has uniform continuous dependence under B perturbations, if for any $\varepsilon > 0$ and any $T > 0$ there exists an $\eta_1(\varepsilon)$, which is independent of T, and an $\eta_2(\varepsilon, T) > 0$ such that $t_0 \geq 0$, $|\varphi(t_0) - y_0| < \eta_1(\varepsilon)$, and $\|h\|_B < \eta_2(\varepsilon, T)$ imply $|\varphi(t) - y(t, t_0, y_0)| < \varepsilon$ on $t_0 \leq t \leq t_0 + T$.

Definition 5. We say that $\varphi(t)$ has uniform finite time attracting under B perturbations, if there exists an $\eta_0 > 0$ and for any $\varepsilon > 0$ there exists a $\tau_1(\varepsilon) > 0$ and an $\eta_3(\varepsilon) > 0$ such that if $t_0 \geq 0$, $|\varphi(t_0) - y_0| < \eta_0$, and $\|h\|_B < \eta_3(\varepsilon)$, then $y(t, t_0, y_0)$ is continuable on $[t_0, t_0 + \tau_1(\varepsilon)]$ and satisfies

$$|\varphi(t_0 + \tau_1(\varepsilon)) - y(t_0 + \tau_1(\varepsilon), t_0, y_0)| < \varepsilon.$$

Theorem 1. If the solution $\varphi(t)$ of (3) has uniform continuous dependence under B perturbations and uniform finite time attracting under B perturbations, then $\varphi(t)$ is BAS.

The proof of BS is analogous to the proof of Theorem 1 in [2]. Letting η_0, η_1, η_2, η_3, and τ_1 be the numbers in Definitions 4 and 5, it is sufficient to set

$$\delta(\varepsilon) = \min\{\rho(\varepsilon), \eta_2(\varepsilon, \tau_1(\rho(\varepsilon))), \eta_3(\rho(\varepsilon))\},$$

where $\rho(\varepsilon) = \min\{\eta_0, \eta_1(\varepsilon)\}$. The rest of the proof follows from uniform finite time attracting under B perturbations by setting $\delta_0 = \eta_0$, $\eta(\varepsilon) = \min\{\delta(\varepsilon), \eta_3(\delta(\varepsilon))\}$, and $\tau(\varepsilon) = \tau_1(\delta(\varepsilon))$. The converse of Theorem 1 is evident from the definitions.

Theorem 2. If the solution $\varphi(t)$ of (3) is IS, then it has uniform continuous dependence under M perturbations.

For a given $\varepsilon > 0$ and $T > 0$, let $\eta_1(\varepsilon) = \delta(\varepsilon)$ and $\eta_2(\varepsilon, T) = \delta(\varepsilon)/(T + 1)$, where $\delta(\cdot)$ is the number in the definition of IS. Then we can see that if $y(t, t_0, y_0)$ is a solution of (4) with $\|h\|_M < \eta_2(\varepsilon, T)$ and if $|\varphi(t_0) - y_0| < \eta_1(\varepsilon)$, we have

$$|\varphi(t) - y(t, t_0, y_0)| < \varepsilon \qquad \text{on} \quad t_0 \le t \le t_0 + T.$$

Theorem 3. If the solution $\varphi(t)$ of (3) is integrally attracting, then it has uniform finite time attracting under M perturbations.

Thus the following theorem is an immediate result from Theorems 1–3.

Theorem 4. If the solution $\varphi(t)$ of (3) is IAS, then it is MAS, and consequently IAS is equivalent to MAS.

As easily seen, Theorems 1–4 are valid for the functional differential equation

$$\dot{x}(t) = f(t, x_t), \tag{5}$$

where $f(t, \varphi)$ is continuous and f takes closed bounded sets of $I \times C([-r, 0], R^n)$ into closed bounded sets of R^n.

Now we shall consider almost-periodic systems or, more generally, systems with the compact hull. For $u \in C(I \times S, R^n)$, where S denotes the space of parameter p in u, let $H(u)$ be the closed hull of u in the compact–open topology, that is, $v \in H(u)$ if and only if there exists a sequence $\{\tau_k\}$, $\tau_k \in I$, such that $u(t + \tau_k, p)$ converges to $v(t, p)$ uniformly on any compact subset of $I \times S$. Clearly, if φ is a solution of (3), for every $(\psi, g) \in H(\varphi, f)$, ψ is a solution of the system

$$x' = g(t, x). \tag{6}$$

Here we should note that if $H(f)$ is compact in the compact–open topology on $I \times S_c$, then for the solution $\varphi(t)$ such that $|\varphi(t)| \le c^* < c$ for all $t \ge 0$, $H(\varphi)$ and consequently $H(\varphi, f)$ are compact. Refer also to [2, Proposition 2]. As before, we assume that system (3) has a bounded solution $\varphi(t)$ such that $\sup_{t \ge 0} |\varphi(t)| \le c^* < c$.

Definition 6. The solution $\varphi(t)$ is asymptotically stable in the hull under B perturbations (called BASH) or the solutions in $H(\varphi)$ are BAS with common $(\delta(\cdot), \delta_0, \tau(\cdot), \eta(\cdot))$, if for every $(\psi, g) \in H(\varphi, f)$, $\psi(t)$ is BAS and the numbers $(\delta(\cdot), \delta_0, \tau(\cdot), \eta(\cdot))$ in the definition of BAS can be chosen independently of (ψ, g).

Theorem 5. Assume that $H(f)$ is compact. Then if the solution $\varphi(t)$ of (3) is uniformly stable in the hull, it has uniform continuous dependence under M perturbations. Moreover, if the solution $\varphi(t)$ of (3) is UASH, it has uniform finite time attracting under M perturbations. Thus, if $H(f)$ is compact, UASH $\Rightarrow$ MAS.

The first part of the theorem corresponds to Lemma 2 and the second part to Lemma 3 in [2]. This theorem can be proved by the same arguments as in the proofs of those lemmas by noting that if $z_k(t)$ is a solution of

$$x' = f(t + \tau_k, x) + h_k(t + \tau_k),$$

and if

$$z_k(0) \to z_0, \qquad f(t + \tau_k, x) \to g(t, x), \qquad \|h_k\|_M \to 0,$$

as $k \to \infty$, then there exists a subsequence of $\{z_k(t)\}$ that converges to a solution of (6) through $(0, z_0)$.

Finally, we shall give the following theorem.

Theorem 6. Assume that $H(f)$ is compact. If the solution $\varphi(t)$ of (3) is TAS, then it is TASH, and consequently it is UASH. If $\varphi(t)$ is MAS, then it is MASH, and consequently it is IASH and also TASH.

If $\varphi(t)$ is TAS, we can show that $\varphi(t)$ is totally stable in the hull, that is, for any $\varepsilon > 0$, any $t_0 \geq 0$, any $(\psi, g) \in H(\varphi, f)$, and any $h \in T$, if

$$|\psi(t_0) - y(t_0)| < \tfrac{1}{2}\delta(\tfrac{1}{2}\varepsilon) \qquad \text{and} \qquad \|h\|_T < \tfrac{1}{2}\delta(\tfrac{1}{2}\varepsilon),$$

then $|\psi(t) - y(t)| < \varepsilon$ for all $t \geq t_0$, whenever $y(t)$ is a solution of

$$y' = g(t, y) + h(t), \tag{7}$$

where $\delta(\cdot)$ is the number for TS of $\varphi(t)$ and we can assume that $\varepsilon < (c - c^*)/2$. If we let

$$\delta_0{}^* = \min\{\tfrac{1}{2}\delta_0, \tfrac{1}{2}\delta((c - c^*)/4)\} \quad \text{and} \quad \eta^*(\varepsilon) = \min\{\tfrac{1}{2}\eta(\tfrac{1}{2}\varepsilon), \tfrac{1}{2}\delta((c - c^*)/4)\},$$

then we can see that $|\psi(t) - y(t)| < \varepsilon$ for all $t \geq t_0 + \tau(\tfrac{1}{2}\varepsilon)$ if $|\psi(t_0) - y(t_0)| < \delta_0{}^*$ and $\|h\|_T < \eta^*(\varepsilon)$, where $\eta(\cdot)$ and $\tau(\cdot)$ are the numbers for TAS.

Thus if $H(f)$ is compact, the following relation is an immediate result of Theorems 4–6:

$$\text{IAS} \Leftrightarrow \text{MAS} \Leftrightarrow \text{TAS} \to \text{UAS},$$
$$\Updownarrow \qquad\qquad\qquad\qquad\qquad (8)$$
$$\text{IASH} \Leftrightarrow \text{MASH} \Leftrightarrow \text{TASH} \Leftrightarrow \text{UASH}.$$

Corollary. If system (3) is a periodic system or if system (3) is an almost-periodic system and for each equation in the hull solutions are unique for initial conditions, then we have relation (8) with $\text{UAS} \Leftrightarrow \text{UASH}$.

Notice that under the assumption in the corollary, UAS implies UASH (see [2, 4]).

Remark 1. Theorem 6 shows that TAS and MAS are inherited.

Remark 2. Uniformly asymptotic stability does not necessarily imply integrally asymptotic stability. See the example in [2].

Theorems 5 and 6 can be extended to functional differential equation (5) by the same arguments as used in the proofs of Lemmas 2 and 3 in [2], where $f(t, \varphi)$ in (5) is bounded on $I \times \overline{C}_c$ and uniformly continuous in (t, φ) on $I \times S$ for any compact set S in $\overline{C}_c$, $C_c = \{\varphi \in C;\ \|\varphi\| < c\}$.

REFERENCES

[1] Shui-Nee Chow and J. A. Yorke, Lyapunov theory and perturbation of stable and asymptotically stable systems, *J. Differential Equations* **15** (1974), 308–321.

[2] J. Kato, Uniformly asymptotic stability and total stability, *Tohoku Math. J.* **22** (1970), 254–269.

[3] J. Kato and T. Yoshizawa, A relationship between uniformly asymptotic stability and total stability, *Funkcial. Ekvac.* **12** (1969), 233–238.

[4] T. Yoshizawa, Asymptotically almost periodic solutions of an almost periodic system, *Funkcial. Ekvac.* **12** (1969), 23–40.

On a General Type of Second-Order Forced Nonlinear Oscillations

R. REISSIG
Institut für Mathematik
Ruhr-Universität, Bochum, Germany

The forced oscillation problem for the generalized Liénard equation (see [4])

$$x'' + f(x)x' + g(x) = p(t) \equiv p(t + T), \qquad \omega = 2\pi/T \qquad (1)$$

(f, g, p continuous functions) has been studied extensively in some recent papers based on the Leray–Schauder fixed-point principle. Assuming

$$\int_0^T p(t)\, dt = 0, \qquad g(x)/x \to 0+ \quad (|x| \to \infty),$$

and

$$f(x) \text{ constant} \qquad \text{or even} \qquad f(x) \text{ arbitrary},$$

Lazer [1] and Mawhin [2] proved the existence of T-periodic solutions. Furthermore, Mawhin considered an analogous vector differential equation.

The purpose of this paper is to replace the rather restrictive condition concerning the restoring term $g(x)$ by the quite natural assumption that this nonlinearity is within the resonance angle if $|x|$ is sufficiently large. Of course, the condition that the mean value of the forcing term must be equal to zero becomes superfluous.

First we complete the results of Lazer [1], Mawhin [2], and others (see [3]) by an obvious extension.

Corollary. If the forcing term has a vanishing mean value and if

$$xg(x) \leq 0 \qquad \text{either for} \quad |x| \geq H \quad \text{or} \quad |x| \leq 0.6T^2 \operatorname{Max}|p(t)|,$$

or

$$0 \leq xg(x) \leq ax^2, \qquad |x| \geq H, \quad a > 0 \text{ small enough},$$

then Eq. (1) admits at least one T-periodic solution. The same is true if $p(t)$ is arbitrary and

$$\lim_{|x| \to \infty} g(x)/x = k \in (0, \omega^2).$$

223

Next we state a more general existence theorem.

Theorem. Equation (1) possesses a T-periodic solution if

$$0 < h \leq g(x)/x \leq k < \omega^2, \qquad |x| \geq H.$$

Proof. Let us consider the auxiliary equation

$$x'' + (1 - \lambda)kx = \lambda\{p(t) - g(x) - f(x)x'\}, \qquad 0 \leq \lambda \leq 1. \tag{2}$$

Its periodic solutions can be represented as $x(t) = \lambda\mathfrak{A}\{x(t)\}$, $\mathfrak{A}$ a completely continuous transformation of the Banach space $\{x(t) \in C^0[0, T] : x(0) = x(T)\}$ supplied with the supremum norm. According to Leray–Schauder, there exists at least one fixed point of $\mathfrak{A}$ [which is a periodic solution of Eq. (1)] provided that the fixed points of $\lambda\mathfrak{A}$, $0 \leq \lambda < 1$, are uniformly bounded. The a priori boundedness is an immediate consequence of some simple statements concerning the oscillatory behavior of the solutions of Eq. (2).

Let $x(t)$, $a \leq t \leq b$ be such a solution satisfying the boundary conditions

$$x(t)\Big|_a^b = 0, \qquad x(t)x'(t)\Big|_a^b = 0. \tag{3}$$

Multiplying Eq. (2) by $x(t)$ and integrating over $[a, b]$ we obtain an estimate

$$\|x'\|_{a,b}^2 \leq k\|x\|_{a,b}^2 + (b - a)^{1/2}M\|x\|_{a,b}, \tag{4}$$

where $\|x\|_{a,b}$ means the L^2-norm of $x(t)$, $a \leq t \leq b$, and M denotes an appropriate constant depending on $g(x)$ and $p(t)$.

First special case: $x(a) = x(b) = 0$. Evidently, in this case

$$\|x\|_{a,b} \leq (b - a)/\pi\|x'\|_{a,b}.$$

Hence

$$\{(\pi/(b - a))^2 - k\}\|x\|_{a,b} \leq (b - a)^{1/2}M.$$

In case $b - a < \pi k^{-1/2}$, uniform bounds are attainable for $\|x\|_{a,b}$, $\|x'\|_{a,b}$. as well as for

$$x(t) = \int_a^t x'(s)\,ds \qquad \text{and} \qquad x'(t) = \int_{a'}^t x''(s)\,ds.$$

Second special case: $x(t) \geq 0 \ (\leq 0)$ on $[a, b]$. Introducing

$$x(\tau) = (b - a)^{-1}\int_a^b x(t)\,dt,$$

we find by means of integration of Eq. (2),

$$|x(\tau)| \leqq h^{-1}M + h^{-1}(b-a)^{-1}|x'(b) - x'(a)|.$$

This estimate is taken into account when the following well-known connection between the L^2-norms of x and x' is evaluated:

$$\|x\|_{a,b}^2 \leqq (b-a)x^2(\tau) + ((b-a)/2\pi)^2 \|x'\|_{a,b}^2.$$

Assuming

$$0 < l \leqq b - a < 2\pi k^{-1/2}$$

and having regard to (4), we may deduce uniform bounds for $\|x\|_{a,b}$, $\|x'\|_{a,b}$, and $x(t) = x(\tau) + \int_\tau^t x'(s)\,ds$, which are, however, linear functions of $|x'(b) - x'(a)|$.

Finally, let $x(t)$ be a T-periodic solution of Eq. (2). If permanently $x(t) \geqq 0$ ($\leqq 0$), then $a = 0$, $b = T$ $[2\pi/(b-a) = \omega]$ is chosen and the second result with $x'(b) - x'(a) = 0$ is applied.

The remaining solutions $x(t)$ have at least three zeros within each compact interval of length T where $x = 0$ at the endpoints. Consider such an interval $[t_0, t_0 + T]$.

Suppose that it can be subdivided into at most three intervals also bounded by zeros and having lengths that do not exceed $T/2$. The first result holds for each of these subintervals.

Suppose on the contrary that there is a subinterval $[a, b]$, $t_0 < a < b \leqq t_0 + T$ with length $b - a > T/2$ and with $x = 0$ only at the endpoints. The first result is applied to $[t_0, a]$, $[b, a + T]$, which leads to an estimate of $|x'(b) - x'(a)|$. Subsequently, the second result is applied to $[a, b]$, where $l = T/2$.

Note that the indicated results may be extended to the higher-order equation

$$x^{(2n)} + f(x)x' + g(x) = p(t), \qquad n \geqq 1 \text{ integer}.$$

REFERENCES

[1] A. C. Lazer, On Schauder's fixed point theorem and forced second-order nonlinear oscillations, *J. Math. Anal. Appl.* **21** (1968), 421–425.

[2] J. Mawhin, An extension of a theorem of A. C. Lazer on forced nonlinear oscillations, *J. Math. Anal. Appl.* **40** (1972), 20–29.

[3] R. Reissig, Extension of some results concerning the generalized Liénard equation, *Ann. Mat. Pura Appl.* **104** (1975), 269–281.

[4] R. Reissig, G. Sansone, and R. Conti, Qualitative theory of nonlinear differential equations (German). Edizione Cremonese, Rome, 1963, pp. xxxii, 381.

Stability of Periodic Linear Systems and the Geometry of Lie Groups*

ROGER W. BROCKETT

Division of Engineering and Applied Physics
Harvard University, Cambridge, Massachusetts

A number of the results on the stability of periodic linear systems of ordinary differential equations hinge on special cases of the following general question. Let G be a Lie group and g its Lie algrbra. Let k be a subspace of g and let $c \subset k$ be a cone in k. Suppose $\|\cdot\|$ is any norm on the space of Lebesgue measurable functions $F = \{f \mid f : [0, \ T] \to k\}$. Given a subset $\tilde{G}$ in G find the infimum of $\|f\|$ over all $f \in F$ with the property that $G(T) \in \tilde{G}$ for

$$\dot{G}(t) = f(t)G(t), \qquad G(0) = \text{identity}, \quad \cdot = d/dt.$$

This kind of variational problem on a Lie group has apparently never been studied explicitly but examples occur frequently in various contexts. Special features that make the question interesting include the fact that k is just a subspace, and not a subalgebra, of g and the cone condition, which gives the problem a semigroup flavor.

It is known that given a matrix equation

$$\dot{X}(t) = \left(\sum_{i=1}^{m} u_i(t)A_i \right) X(t), \qquad X(0) = I,$$

there exists a set of functions $\{u_i(\cdot)\}$ that steer the system to X_1 if and only if $X_1 \varepsilon \{\exp\{A_i\}_{LA}\}_G$; that is, X_1 must belong to the smallest group of matrices that contains $\exp\{A_i\}_{LA}$, where $\{A_i\}_{LA}$ indicates the Lie algebra generated by A_i (see [1]). If the u_i are restricted to lie in a cone then the problem becomes more complex, and less satisfactory resolutions are available. However, if there exists a control that steers X to a given desired state then we can inquire about controls that accomplish the same task and are optimal with respect to some criterion.

In their beautiful paper on the stability of linear periodic canonical

---

* This research was supported by the U.S. Office of Naval Research under the Joint Services Electronics Program by Contract N00014-75-C-0648.

systems, Gel'fand and Lidskiĭ [2] investigate certain aspects of the geometry of the group of symplectic matrices. They show that the interior of the subset of the $2n \times 2n$ symplectic matrices $(Sp(n))$, which consists of those matrices similar to orthogonal ones, has 2^n connected components—a particular component being characterized by the distribution of Floquet multipliers of first and second type in the sense of Kreĭn [3]. Kreĭn's work (see [4] for a survey) establishes that for Hill's equation one of these connected components plays a particularly important role and he was able to generalize the classical result of Liapunov on the stability of $\ddot{x}(t) + p(t)x(t) = 0$ in an elegant way. Our purpose here is to consider further the geometric content of the Liapunov–Kreĭn theorems in the context of the theory of Lie groups. The idea is to observe that many results on the stability of Hill's equation have an interpretation in terms of a variational problem of the above type on the symplectic group. At the same time, we have observed earlier [5] that boundedness, asymptotic stability, etc., are properties of linear systems of differential equations that are preserved under tensoring and reduction. Since the tensor product of two symplectic systems yields an orthogonal system we are able to extend further the scope of these results.

Theorem 1. Let α and β be Lebesgue measurable and nonnegative functions defined on $[0, T]$. If $X(t) \in Sp(1)$ is the unique absolutely continuous solution of

$$\dot{X}(t) = \begin{bmatrix} 0 & \alpha(t) \\ -\beta(t) & 0 \end{bmatrix} X(t), \qquad X(0) = I, \tag{†}$$

and if -1 is an eigenvalue of $X(T)$, then

$$\frac{1}{2} \int_0^T \alpha(t) + \beta(t)\, dt \geq \left[\int_0^T \alpha(t)\, dt \int_0^T \beta(t)\, dt \right]^{1/2} \geq 2, \tag{*}$$

Proof. If -1 is an eigenvalue of $X(T)$ then there is some vector x_0 such that $X(T)x_0 = -x_0$. Moreover, $X(t)x_0$ satisfies the vector version of the given equation. Describing the vector equation in polar coordinates gives

$$\dot{r}(t) = (\alpha - \beta)r(t) \cos \theta(t) \sin \theta(t),$$

$$\dot{\theta}(t) = \alpha(t) \cos^2 \theta(t) + \beta(t) \sin^2 \theta(t) = \langle [\alpha(t), \beta(t)], [\cos^2 \theta, \sin^2 \theta] \rangle.$$

From the second equation we see, after a calculation, that to increase θ from $\theta(0)$ to $\theta(0) + \pi$ we must have

$$\int_0^T \alpha(t) + \beta(t)\, dt \geq 4.$$

The effect of the similarity transformation

$$X(t) \mapsto \begin{pmatrix} \rho & 0 \\ 0 & 1 \end{pmatrix} X(t) \begin{pmatrix} 1/\rho & 0 \\ 0 & 1 \end{pmatrix}$$

is such as leave all aspects of the problem invariant except $\alpha \to \rho\alpha$ $\beta \to (1/\rho)\beta$. Thus we see that for all $\rho > 0$,

$$\frac{1}{2} \int_0^T \rho\alpha(t) + \frac{1}{\rho} \beta(t) \, dt \geq 2.$$

This fact allows us to reverse the usual arithmetic–geometric mean inequality and to conclude (∗).

Theorem 1 is equivalent to Liapunov's inequality stating that for $x(0) = x(T) = 0$ and $x(t) > 0$ for $t \in (0, T)$,

$$\int_0^T |\ddot{x}/x| \, dt \geq 4/T.$$

Both allow one to conclude easily that $\ddot{x}(t) + p(t)x(t) = 0$ is stable for $p(t + T) = p(t)$ and $\int_0^T p(t) \, dt \leq 4/T$. Thus Theorem 1 captures the content of Liapunov's inequality in terms of the geometry of the group $Sp(1)$ and should be compared with other work in this direction (see [6] and references therein).

Theorem 2. Let $A(t) = A'(t)$ and $B(t) = B'(t)$ be $n \times n$ matrix-valued functions of time whose entries are defined and Lebesgue measurable on $[0, T]$ and are nonnegative definite there. If $X(t) \in Sp(n)$ is the unique absolutely continuous solution of

$$\dot{X}(t) = \begin{bmatrix} 0 & A(t) \\ -B(t) & 0 \end{bmatrix} X(t), \qquad X(0) = I,$$

and if -1 belongs to the sprectrum of $X(T)$, then

$$\frac{1}{2} \int_0^T \lambda_{\max} A(t) + \lambda_{\max} B(t) \, dt \geq \left[\int_0^T \lambda_{\max} A(t) \, dt \int_0^T \lambda_{\max} B(t) \, dt \right]^{1/2} \geq 2. \quad (**)$$

Proof. The given equations can be regarded as the state–costate pair for the variational problem defined by the constraints $\dot{x}(t) = N(t)u(t)$, $x(0) = x(T) = 0$, and the performance functional

$$\eta = \min_{u(\cdot)} \int_0^T u'(t)u(t) - x'(t)B(t)x(t) \, dt,$$

where $N'(t)N(t) = A(t)$ (see, e.g., [7, p. 158]). Now if $A(t)$ is actually positive definite on $[0, T]$ then $u(t) = N^{-1}(t)\dot{x}(t)$ and

$$\eta \geq \int_0^T (\lambda_{\min}[A^{-1}(t)])\dot{x}'(t)\dot{x}(t) - (\lambda_{\max}[B(t)])x'(t)x(t) \, dt,$$

Using the known disconjugacy results for scalar problems and Theorem 1, the integral on the left has a minimum if (**) holds. An appropriate limiting argument omitted here lets us take $A(t)$ nonnegative definite as well. This disconjugacy allows us to conclude that -1 is not in the spectrum of $X(T)$.

Theorem 2, when applied to the stability problem yields the following theorem of Krein.

Theorem 3. Let $A(t) = A'(t)$ and $B(t) = B'(t)$ be positive definite, Lebesgue measurable, and periodic on $(-\infty, \infty)$ with period T. Then the differential equation

$$\dot{X}(t) = \begin{bmatrix} 0 & A(t) \\ -B(t) & 0 \end{bmatrix} X(t)$$

is strongly stable if

$$\left[\int_0^T \lambda_{\max}[A(t)] \, dt \int_0^T \lambda_{\max}[B(t)] \, dt \right]^{1/2} < 2.$$

We mention one further example of this type due to Neĭgauz and Lidskiĭ (see [4, p. 242]).

Theorem 4. Let $M(t) = M'(t)$ be a $2n \times 2n$ matrix-valued function of time whose entries are defined and Lebesgue measurable on $[0, T]$ and which is nonnegative definite there. If $X(t)$ is the unique absolutely continuous solution of

$$\dot{X}(t) = \begin{bmatrix} 0 & I \\ -I & 0 \end{bmatrix} M(t)X(t), \qquad X(0) = I,$$

and if -1 belongs to the spectrum of $X(t)$, then

$$\int_0^T \|M(t)\| \, dt \geq \pi.$$

Incidently, related results imply that if $X(t)$ is to be periodic of period T

for an equation of this type on $Sp(n)$, then

$$\int_0^T \|M(t)\| \, dt \geq 2\pi.$$

We illustrate the related facts on $SO(p, q)$ via an example, omitting generalizations for lack of space. Equation (†) becomes under Kronecker squaring $(z_1 = x^2 x_2,\ z_2 = 2x\dot{x},\ z_3 = \dot{x}^2 + x^2)$

$$\begin{bmatrix} \dot{z}_1(t) \\ \dot{z}_2(t) \\ \dot{z}_3(t) \end{bmatrix} = \begin{bmatrix} 0 & \gamma(t) & 0 \\ -\gamma(t) & 0 & \delta(t) \\ & \delta(t) & 0 \end{bmatrix} \begin{bmatrix} z_1(t) \\ z_2(t) \\ z_3(t) \end{bmatrix}$$

where $\gamma(t) = \alpha(t) + \beta(t)$, $\delta(t) = \alpha(t) - \beta(t)$. We see then that the content of Liapunov's inequality on $SO(2,1)$ is that for any nontrivial periodic solution of this equation with $\gamma(t) \geq 0$ and $|\delta(t)| \leq \gamma(t)$ we have $\int_0^T \gamma(t) \, dt \geq 4$. Generalizations to higher-dimensional versions follow by a systematic use of the representation theory of Lie algebras. Specific results in this direction may be seen in [5, Section 5].

REFERENCES

[1] R. W. Brockett, System theory on group manifolds and coset spaces, *SIAM J. Control* **10** (1972), 265–284.

[2] I. M. Gel'fand and V. B. Lidskiĭ, On the structure of the regions of stability of linear canonical systems of differential equations with periodic coefficients, *Amer. Math. Soc. Transl. Ser. 2* **8** (1958), 143–181.

[3] M. G. Kreĭn, Generalization of certain investigations of A.M. Liapunov on linear differential equations with periodic coefficients, *Dokl. Akad. Nauk. SSSR* **73** (1950), 445–448 (in Russian).

[4] Ju. L. Daleckiĭ and M. G. Kreĭn, "Stability of Solutions of Differential Equations in Banach Space," Vol. 43. Transl. Math. Amer. Math. Soc., 1974.

[5] R. W. Brockett, Lie algebras and lie groups in control theory, in "Geometric Methods in System Theory" (D. Q. Mayne and R. W. Brockett, eds.). Reidel, Dordrecht, Holland, 1973.

[6] H. Guggenheimer, Geometric theory of differential equations I: Second order linear equations, *SIAM J. Math. Anal.* **2** (1971), 233–241.

[7] R. W. Brockett, "Finite Dimensional Linear Systems." Wiley, New York, 1970.

Periodic Solutions of Holomorphic Differential Equations

N. G. LLOYD*

Department of Mathematics
St. John's College, Cambridge, England

Suppose M is a complex manifold; let $\mathscr{F}$ be the collection of holomorphic time-dependent vector fields of (fixed) period ω. So $f \in \mathscr{F}$ when f maps $M \times \mathbb{R}$ continuously into TM, the tangent bundle of M, in such a way that

(1) $f(x, t + \omega) = f(x, t)$,
(2) f is holomorphic in x,
(3) $f(\cdot, t)$ is a section of TM for each t.

$\mathscr{F}$ is given the topology of uniform convergence on compact sets of $M \times \mathbb{R}$. The orbits of f are given locally by the solutions of an ordinary differential equation

$$\dot{z} = f_*(t, z).$$

Let the orbit of f satisfying $z(t_0) = z_0$ be denoted $z_f(t; t_0, z_0)$.

Suppose $\mathscr{S}$ is a closed subset of $\mathscr{F}$. If there are sequences f_n, ξ_n in $\mathscr{S}$, M, respectively, with $f_n \to f$, $\xi_n \to \xi$, and a real number τ such that each $z_{f_n}(t; \tau, \xi_n)$ is of period $m\omega$, then $z_f(t; \tau, \xi)$ is either of period $m\omega$ or is undefined for $0 \le t \le m\omega$. In the latter case, we say that $z_f(t; \tau, \xi)$ is an $m\omega$-singular periodic solution of f relative to $\mathscr{S}$ (written $m\omega$-sps).

Two particular cases are of special interest: (i) M compact, (ii) $M = \mathbb{C}^N$. When M is compact there are no singular periodic solutions, and the difficulties lie elsewhere. Here *we deal with the case $M = \mathbb{C}^N$*; so we consider the family of equations

$$\dot{z} = f(t, z) \qquad (1)$$

with f holomorphic in z and of period ω in t.

Let $q_m^{(\tau)}(f, c) = z_f(m\omega + \tau; \tau, c) - c$; write q_m for $q_m^{(0)}$. If $\varphi(t)$ is an isolated $m\omega$-periodic solution of (1), define the multiplicity of φ to be the

---

* Present address: Department of Mathematics, University of Wales, Aberystwyth, U.K.

local degree of $q_m(f, \cdot)$ at zero relative to $\varphi(0)$. The multiplicity so defined is always at least $+1$.

By using the fact that compact analytic sets in $\mathbb{C}^N$ are finite, it is possible to prove the analog of Rouché's theorem:

Theorem 1. Let D be a bounded domain in $\mathbb{C}^N$ and let f, g be holomorphic maps of $\overline{D}$ into $\mathbb{C}^N$ with $|g| < |f|$ on ∂D. If f has finitely many zeros in D, then f and $f + g$ have the same number of zeros in D, counting multiplicity.

Using this result, the following may be proved.

Theorem 2. Let $\pi_m(f)$ be the number of isolated $m\omega$-periodic solutions of f. If $\pi_m(f) < \infty$, there is a neighborhood U of f such that for $g \in U$, $\pi_m(g) \geq \pi_m(f)$, again counting multiplicity.

For every integer m, it is convenient to define the following sets:

$$\mathscr{B}_m = \{f \in \mathscr{S}; f \text{ has } m\omega\text{-sps}\},$$
$$\mathscr{I}_m = \{f \in \mathscr{S}; f \text{ has infinitely many } m\omega\text{-ps}\},$$
$$\mathscr{A}_m = \mathscr{S} \backslash (\mathscr{B}_m \cup \mathscr{I}_m).$$

As f changes, periodic solutions are lost only when an sps occurs, and that happens when the amplitudes of those periodic solutions become infinitely large. An $m\omega$-sps tends to infinity both as t increases and as t decreases. Formally, we have

Lemma 1. (a) An $m\omega$-sps is defined for a time less than $m\omega$.
(b) If $f \in \mathscr{B}_m$, there are sequences f_n, α_n, and $\tau \in \mathbb{R}$ such that

$$q_m^{(\tau)}(f_n, \alpha_n) = 0, \qquad \alpha_n \to \infty.$$

Singular periodic solutions are comparatively rare, in the sense that it can be shown that $\mathscr{A}_m \cup \mathscr{I}_m$ is dense in $\mathscr{S}$ for all m. The aim is to replace the inequality in Theorem 2 with equality. To do so, however, it is necessary to introduce a condition on the equations considered; this condition has the effect of controlling the behavior of solutions near infinity.

Hypothesis A. There is a function $\rho: \mathscr{S} \to \mathbb{R}$ such that every $m\omega$-ps of f meets the ball $B(0, \rho(f))$, and every f has a neighborhood U_f in $\mathscr{S}$ such that $\rho[U_f]$ is bounded.

Suppose $\mathscr{S}$ satisfies this hypothesis; we then have the following results.

Lemma 2. (a) The converse of Lemma 1 (b) holds.

(b) $\mathscr{I}_m \subset \mathscr{B}_m$ (for every m).

(c) $\mathscr{A}_m$ is open (for every m).

We can now obtain the result we have sought.

Theorem 3. Let $\mathscr{S}$ satisfy Hypothesis A and let $f \in \mathscr{A}_m$. There is a neighborhood U of f such that $\pi_m(g) = \pi_m(f)$ for all $g \in U$.

Corollary. For $f \in \mathscr{A}_m$, let $C_m(f)$ be the component of f in $\mathscr{A}_m$. For $g \in C_m(f)$, $\pi_\lambda(g) = \pi_\lambda(f)$ for $\lambda \mid m$.

Remark. Both the theorem and the corollary take account of multiplicity.

We give the proof of the theorem. First $f \notin \mathscr{I}_m$. From Theorem 2, $\pi_m(g) \geq \pi_m(f) = M$ if g is near enough to f. If $f_n \to f$, $\pi_m(f_n) > M$, there are $m\omega$-ps ψ_n of f_n bounded away from the $m\omega$-ps of f (again by Theorem 2). So there is a subsequence of $\{\psi_n(0)\}$ that tends either to infinity or to a point γ with $z_f(t; 0, \gamma)$ not periodic. Thus $f \in \mathscr{B}_m$ in both cases—by Lemma 2(a) in the former case, by definition in the latter. But it was supposed that $f \in \mathscr{A}_m$; hence all g near enough to f have exactly M $m\omega$-ps.

On the Newton Method of Solving Problems of the Least Squares Type for Ordinary Differential Equations

MINORU URABE

Department of Mathematics, Faculty of Science
Kyushu University, Fukuoka, Japan

1

Consider the differential equation

$$dx/dt = X(x, t), \qquad (1)$$

where x and $X(x, t)$ are real n-dimensional vectors and $X(x, t)$ is defined and twice continuously differentiable with respect to x in the region D of the tx-space intercepted by two hyperplanes $t = a$ and $t = b$ $(a < b)$. In this paper, we present a method of obtaining a solution $x = x(t)$ of (1) on the interval $I = [a, b]$ such that

$$\sum_{i=1}^{M} \left\| \sum_{j=1}^{N} L_{ij} x(t_j) - l_i \right\|^2 = \min, \qquad (2)$$

where L_{ij} are the given $n \times n$ constant square matrices, t_j the given points on the interval I, l_i the given n-dimensional constant vectors, and the symbol $\|\cdot\|$ denotes the Euclidean norm of vectors.

An iterative method of obtaining a solution to the above problem has been given recently by Banks and Groome [1] by the use of the quasi-linearization of differential equation (1). However, as will be shown in Section 2, the problem is reduced to the solution of a boundary value problem with a nonlinear boundary condition. For such boundary value problems, the author showed in his paper [3] that the Newton method can be used effectively for obtaining a desired solution. Here, we will show that the method established in [3] can be readily applied to the present problem and that by doing so one obtains two methods, one more efficient and the other simpler compared with the method proposed by Banks and Groome in their paper [3].

2

In this section, we shall show that the problem of finding a solution of (1) satisfying condition (2) is reduced to a boundary value problem with a particular nonlinear boundary condition.

Write an arbitrary solution of (1) existing on I and lying in D as $x = x(t, c)$, where c is a value of the solution at an arbitrary fixed point $t = t_0 \in I$. Consider the function $J(c)$ defined by

$$J(c) = \frac{1}{2} \sum_{i=1}^{M} \left\| \sum_{j=1}^{N} L_{ij} x(t_j, c) - l_i \right\|^2. \tag{3}$$

Then for the gradient vector $J'(c)$ of $J(c)$ with respect to c, we readily have

$$J'(c) = \sum_{i=1}^{M} \left[\sum_{j=1}^{N} L_{ij} x_c(t_j, c) \right]^* \left[\sum_{j=1}^{N} L_{ij} x(t_j, c) - l_i \right], \tag{4}$$

where $x_c(t, c)$ denotes the Jacobian matrix of $x(t, c)$ with respect to c and the symbol $*$ denotes transpose of vectors or matrices. Clearly the extremal value of $J(c)$ is attained at $c = c_0$ satisfying the equation

$$J'(c) = 0, \tag{5}$$

and conversely, as is seen easily, $c = c_0$ satisfying (5) really minimizes $J(c)$ locally if the matrix $\sum_{j=1}^{N} L_{ij} x_c(t_j, c_o)$ is nonsingular for at least one i and

$$\left\| \sum_{j=1}^{N} L_{ij} x(t_j, c_0) - l_i \right\|$$

are sufficiently small for all i. The latter condition means that the solution $x = x(t, c_0)$ satisfies the boundary conditions

$$\sum_{j=1}^{N} L_{ij} x(t_j) - l_i = 0, \qquad i = 1, 2, \ldots, M, \tag{6}$$

approximately and, by [2], the former condition means that the solution $x = x(t, c_0)$ is isolated for some of the boundary conditions (6). Both conditions are natural requisites for our problem, which is thus reduced to the problem of finding $c = c_0$ satisfying Eq. (5). However, $x_c(t, c)$ is clearly a fundamental matrix of the first variation equation of (1) with respect to the solution $x(t, c)$. Therefore, as is seen from (4), our problem is reduced to the problem of finding a solution $x = x(t)$ of (1) satisfying the nonlinear boundary condition as follows:

$$f(x) := \sum_{i=1}^{M} \left[\sum_{j=1}^{N} L_{ij} \Phi_{(x)}(t_j) \right]^* \left[\sum_{j=1}^{N} L_{ij} x(t_j) - l_i \right] = 0, \tag{7}$$

where $\Phi_{(x)}(t)$ is the fundamental matrix of the linear homogeneous differential equation

$$dy/dt = X_x[x(t), t]y \tag{8}$$

satisfying the initial condition $\Phi_{(x)}(t_0) = E$ for a certain fixed $t_0 \in I$. In (8), $X_x(x, t)$ denotes the Jacobian matrix of $X(x, t)$ with respect to x. In what follows, the problem of finding a solution of (1) satisfying boundary condition (7) will be called the problem (P).

3

In this section, we shall give an iterative procedure for our problem obtained by applying the Newton method established in [2] to the problem (P).

The iterative procedure obtained by which a solution of (1) satisfying (2) is successively calculated is

$$x^{(p+1)}(t) = \zeta_{(x^{(p)})}(t) - \Phi_{(x^{(p)})}(t)[Q(x^{(p)}) + R(x^{(p)})]^{-1}$$
$$\times [\rho(x^{(p)}) + \sigma(x^{(p)})], \qquad p = 0, 1, 2, \ldots, \tag{9}$$

where

$$\zeta_{(x)}(t) = \Phi_{(x)}(t)x(t_0) + \Phi_{(x)}(t) \int_0^t \Phi_{(x)}^{-1}(s)$$
$$\times \{X[x(s), s] - X_x[x(s), s]x(s)\} \, ds,$$

$$Q(x) = \sum_{i=1}^{M} \left[\sum_{j=1}^{N} L_{ij} \Phi_{(x)}(t_j) \right]^* \left[\sum_{j=1}^{N} L_{ij} \Phi_{(x)}(t_j) \right],$$

$$R(x) = \sum_{i=1}^{M} \left[\sum_{j=1}^{N} L_{ij} x(t_j) - l_i \right]^* \left[\sum_{j=1}^{N} L_{ij} \{\Psi_{(x)} \Phi_{(x)}\}(t_j) \right],$$

$$\rho(x) := \sum_{i=1}^{M} \left[\sum_{j=1}^{N} L_{ij} \Phi_{(x)}(t_j) \right]^* \left[\sum_{j=1}^{N} L_{ij} \zeta_{(x)}(t_j) - l_i \right],$$

$$\sigma(x) = \sum_{i=1}^{M} \left[\sum_{j=1}^{N} (\Psi_{(x)}(\zeta_{(x)} - x)(t_j) \right]^* \left[\sum_{j=1}^{N} L_{ij} x(t_j) - l_i \right],$$

and $\Psi_{(x)}(t)$ is the third-order tensor defined by

$$[\Psi_{(x)}h](t) = \Phi_{(x)}(t)\int_{t_0}^{t}\Phi_{(x)}^{-1}(s)X_{xx}[x(s), s]h(s)\Phi_{(x)}(s)\,ds, \tag{10}$$

for an arbitrary n-dimensional vector-valued function $h(t)$ continuous on I. In (10), $X_{xx}(x, t)$ denotes the third-order tensor whose components are the second-order derivatives of $X(x, t)$ with respect to x.

The iterative procedure (9) is the Newton iterative procedure applied to the problem (P); therefore it is quadratically convergent if the starting approximate solution $x^{(0)}(t)$ is sufficiently accurate and is isolated in the sense that the matrix $\sum_{j=1}^{N} L_{ij}\Phi_{(x^{(0)})}(t_j)$ is nonsingular for at least one i.

If we apply the generalized Newton method to the problem (P) replacing the derivatives appearing in the Newton method by some suitable functions, then we have various linearly convergent iterative procedures, among which the typical ones are

$$x^{(p+1)}(t) = \tilde{\zeta}_{(x^{(p)})}(t) - \Phi(t)\tilde{Q}^{-1}\tilde{\rho}(x^{(p)}), \tag{11}$$

$$x^{(p+1)}(t) = \zeta_{(x^{(p)})}(t) - \Phi_{(x^{(p)})}(t)Q^{-1}(x^{(p)})\rho(x^{(p)}). \tag{12}$$

In (11), $\Phi(t)$ is the fundamental matrix of a linear homogeneous differential equation

$$dy/dt = A(t)y$$

satisfying the initial condition $\Phi(t_0) = E$, where $A(t)$ is a continuous square matrix close to $X_x[x^{(0)}(t), t]$; $\tilde{\zeta}_{(x)}(t)$ is a vector-valued function obtained from $\zeta_{(x)}(t)$ by replacing $\Phi_{(x)}(t)$ and $X_x[x(s), s]$ by $\Phi(t)$ and $A(s)$, respectively; $\tilde{Q}$ is a constant square matrix obtained from $Q(x)$ by replacing $\Phi_{(x)}(t)$ by $\Phi(t)$; and

$$\tilde{\rho}(x) = \sum_{i=1}^{M}\left[\sum_{j=1}^{N}L_{ij}\Phi(t_j)\right]^{*}\left[\sum_{j=1}^{N}L_{ij}\{\tilde{\zeta}_{(x)} - x\}(t_j)\right]$$

$$+ \sum_{i=1}^{M}\left[\sum_{j=1}^{N}L_{ij}\Phi_{(x)}(t_j)\right]^{*}\left[\sum_{j=1}^{N}L_{ij}x(t_j) - l_i\right]. \tag{13}$$

The iterative procedure (12) is the same one established by Banks and Groome [1].

REFERENCES

[1] H. T. Banks and G. M. Groome, Jr., Convergence theorems for parameter estimation by quasilinearization, *J. Math. Anal. Appl.* **42** (1973), 91–109.

[2] M. Urabe, An existence theorem for multipoint boundary value problems, *Funkcial. Ekvac.* **9** (1966), 43–60.

[3] M. Urabe, The Newton method and its application to boundary value problems with nonlinear boundary conditions, *Proc. U.S.–Japan Seminar Differential Functional Equations*, pp. 383–410. Benjamin, New York, 1967.

A Study on Generation of Nonuniqueness*

YASUTAKA SIBUYA
School of Mathematics
University of Minnesota, Minneapolis, Minnesota

Introduction

Under certain assumptions on stability of a bounded solution, it can be shown that a given almost-periodic system has an almost-periodic solution (cf. Yoshizawa [3, 4].) In such a study, some differences have been noticed between the two cases:

(i) the case when every equation in the hull satisfies a uniqueness condition of solutions of initial-value problems, and

(ii) the case when some equations in the hull do not satisfy such conditions.

Recently, Sell [2] has proposed a new method that is applicable to both of these cases. Through an examination of Sell's method, it was noticed that differences between cases (i) and (ii) are largely due to the following fact: Let us consider a system

$$dx/dt = f(t, x), \qquad (S)$$

and its perturbation

$$dy/dt = f(t, y) + g(t, y). \qquad (S')$$

If (S) satisfies a uniqueness condition, every solution of (S) can be approximated by solutions of (S'). However, if (S) does not satisfy such a condition, there may exist a solution of (S) that cannot be approximated by any solutions of (S'). Keeping this fact in mind, we shall investigate the generation of nonuniqueness through perturbations. Particular attention will be paid to those perturbations that preserve important properties of solutions such as boundedness, periodicity, almost-periodicity, non-almost-periodicity, and various stability properties. As an example of applications, a differential equation with almost-periodic coefficients will be constructed

* This research was partially supported by NSF Grant GP-38955.

so as to show clearly that uniform asymptotic stability need not be inherited by the solutions of differential equations in the hull. Such an example was given also by Kato [1]. Our example will be constructed more systematically.

1. Preliminary Observations

Consider a family of curves

$$x = \sin (t - c) + c, \tag{1.1}$$

where c is a real parameter. We shall show that the differential equation for family (1.1) admits infinitely many singular solutions

$$x = t + 2n\pi, \qquad n \in Z, \tag{1.2}$$

where Z is the set of all integers. By eliminating c, we derive the differential equation for family (1.1). Note that (1.1) implies

$$x - t = \sin (t - c) - (t - c), \qquad dx/dt = \cos (t - c). \tag{1.3}$$

Let

$$u = G(v) - v \tag{1.4}$$

be the inverse of

$$v = \sin u - u. \tag{1.5}$$

Then $G(v)$ is continuous in v, $G(v)$ is periodic of period 2π in v, and $G(2n\pi) = 0$ for $n \in Z$. Furthermore, $G(v)$ is smooth except at $v = 2n\pi$ ($n \in Z$). The differential equation for (1.1) is

$$dx/dt = \cos (G(x - t) - (x - t)). \tag{1.6}$$

It is easy to prove that (1.2) satisfies (1.6).

Consider next a family of curves

$$x = \varepsilon \sin (t - c) + c, \tag{1.7}$$

where ε is a real parameter such that $0 \leq \varepsilon \leq 1$. Let

$$u = g(v, \varepsilon) - v \tag{1.8}$$

be the inverse of

$$v = \varepsilon \sin u - u. \tag{1.9}$$

Then $g(v, \varepsilon)$ is continuous in v and ε for $-\infty < t < +\infty$, $0 \leq \varepsilon \leq 1$; $g(v, \varepsilon)$ is continuously differentiable with respect to v if $0 \leq \varepsilon < 1$; $g(v, 1) = G(v)$, $g(v, 0) = 0$; $g(v, \varepsilon)$ is periodic of period 2π in v for each fixed ε; and $g(2n\pi, \varepsilon) = 0$ for $n \in Z$, $0 \leq \varepsilon \leq 1$. The differential equation for family (1.7) is

$$dx/dt = \varepsilon \cos \left(g(x - t, \varepsilon) - (x - t)\right). \tag{1.10}$$

As long as $0 \leq \varepsilon < 1$, differential equation (1.10) satisfies a uniqueness condition. At $\varepsilon = 1$, differential equation (1.10) becomes differential equation (1.6). Family (1.7) is uniformly stable in the sense of Liapunov if $0 \leq \varepsilon < 1$. However, at $\varepsilon = 1$, singular solutions (1.2) appear. Therefore, solutions of (1.6) become unstable.

2. Perturbation of a Differential Equation

Consider a differential equation

$$dw/dt = f(t, w) \tag{2.1}$$

and the transformation

$$x = \varepsilon \sin (t - w) + w, \tag{2.2}$$

where w is a real unknown quantity and ε a real parameter such that $0 \leq \varepsilon \leq 1$. The inverse of (2.2) is

$$w = x - g(x - t, \varepsilon), \tag{2.3}$$

and the differential equation for x is given by

$$dx/dt = \varepsilon \cos \left(g(x - t, \varepsilon) - (x - t)\right)\{1 - f(t, x - g(x - t, \varepsilon))\} \tag{2.4}$$
$$+ f(t, x - g(x - t, \varepsilon)).$$

If f is smooth in w, differential equation (2.4) satisfies a uniqueness condition for $0 \leq \varepsilon < 1$. At $\varepsilon = 1$, (2.4) becomes

$$dx/dt = \cos \left(G(x - t) - (x - t)\right)\{1 - f(t, x - G(x - t))\}$$
$$+ f(t, x - G(x - t)),$$

and functions (1.2) satisfy (2.5). This means that the uniqueness condition of differential equation (2.4) breaks down at $\varepsilon = 1$. Note that at $\varepsilon = 0$ transformation (2.2) is the identity transformation $x = w$. Furthermore, (2.2) preserves important properties of solutions such as those listed in the introduction. Periodicity of period 2π and almost-periodicity of the differential equation is also preserved. At $\varepsilon = 1$, solutions of differential equation (2.4) become highly unstable because of singular solutions (1.2).

3. Construction of a Differential Equation with Almost-Periodic Coefficients

Consider a differential equation

$$dw/dt = f(t, w),$$ (3.1)

and a transformation

$$x = a(t) \sin (t - w) + w,$$ (3.2)

where w is a real unknown quantity and $a(t)$ a function of t. If $0 \leq a(t) \leq 1$, the inverse of (3.2) is

$$w = x - g(x - t, a(t)).$$ (3.3)

Assume that $a(t)$ and $da(t)/dt$ are continuous in t for $-\infty < t < +\infty$. Then the differential equation for x is given by

$$\frac{dx}{dt} = a(t) \cos (g(x - t, a(t)) - (x - t))\{1 - f(t, x - g(x - t, a(t)))\}$$

$$+ f(t, x - g(x - t, a(t))) + \frac{da(t)}{dt} \sin (g(x - t, a(t)) - (x - t)).$$ (3.4)

If f is smooth in w, differential equation (3.4) satisfies a uniqueness condition as long as $0 \leq a(t) < 1$. If $a(t) = 1$ for $t \in I = [\tau_1, \tau_2]$, then (3.4) becomes

$$dx/dt = \cos(G(x - t) - (x - t))\{1 - f(t, x - G(x - t))\}$$
$$+ f(t, x - G(x - t))$$ (3.5)

in the interval I, and functions (1.2) become singular solutions of (3.4) there.

It is known that there exists a function $a(t)$ such that

  (i) $a(t)$ and $da(t)/dt$ are continuous in t for $-\infty < t < +\infty$;
  (ii) $a(t)$ and $da(t)/dt$ are almost-periodic in t;
  (iii) $0 \leq a(t) < 1$ for $-\infty < t < +\infty$;

(iv)   There exist two increasing sequences $\{n_k\}$ and $\{m_j\}$ of integers such that

$$a^*(t) = \lim_{k \to +\infty} a(t + 2n_k \pi) \quad \text{and} \quad da^*(t)/dt = \lim_{k \to +\infty} da(t + 2n_k \pi)/dt \tag{3.6}$$

exist uniformly in t, and

$$a^*(t) = 1 \quad \text{for} \quad |t - 2m_j \pi| \leqq \pi, \quad j = 1, 2, \dots. \tag{3.7}$$

Let $f(t, w)$ be continuously differentiable in w and almost-periodic in t uniformly for w. Then (3.4) satisfies a uniqueness condition, and the right-hand side of (3.4) is almost-periodic in t uniformly for w. Assume that

$$f^*(t, w) = \lim_{k \to +\infty} f(t + 2n_k \pi, w) \tag{3.8}$$

uniformly in (t, w). Then the differential equation

$$\frac{dx}{dt} = a^*(t) \cos \left(g(x - t, a^*(t)) - (x - t)\right)\{1 - f^*(t, x - g(x - t, a^*(t)))\}$$

$$+ f^*(t, x - g(x - t, a^*(t))) + \frac{da^*(t)}{dt} \sin \left(g(x - t, a^*(t)) - (x - t)\right) \tag{3.9}$$

belongs to the hull of (3.4). Since $a^*(t)$ satisfies condition (3.7), differential equation (3.9) admits singular solutions (1.2) in the intervals

$$|t - 2m_j \pi| \leqq \pi, \quad j = 1, 2, \dots. \tag{3.10}$$

By choosing $f(t, w)$ in suitable ways, we can construct examples showing that various stability properties need not be inherited by the solutions of differential equations in the hull. In particular, if we choose $f(t, w) = -w$, we can show that uniform asymptotic stability need not be inherited. In this case, (3.4) becomes

$$\frac{dx}{dt} = a(t) \cos \left(g(x - t, a(t)) - (x - t)\right)\{1 + x - g(x - t, a(t))\}$$

$$- \{x - g(x - t, a(t))\} + \frac{da(t)}{dt} \sin \left(g(x - t, a(t)) - (x - t)\right). \tag{3.11}$$

The solution

$$x = a(t) \sin t \tag{3.12}$$

is uniformly asymptotically stable in the sense of Liapunov as $t \to +\infty$.

REFERENCES

[1] J. Kato, Uniformly asymptotic stability and total stability, *Tôhoku Math. J.* **22** (1970), 254–269.

[2] G. R. Sell, Differential equations without uniqueness and classical topological dynamics, *J. Differential Equations* **14** (1973), 42–56.

[3] T. Yoshizawa, Stability for almost periodic systems, *Japan–U.S. Seminar Ord. Differential Functional Equations*, Springer Lecture Notes No. 243, pp. 29–39 (1971).

[4] T. Yoshizawa, Stability theory and the existence of periodic solutions and almost periodic solutions, *Appl. Math. Sci.* **14**, Springer-Verlag, Berlin and New York, 1975.

Relative Asymptotic Equivalence with Weight t^μ, between Two Systems of Ordinary Differential Equations

HILDEBRANDO MUNHOZ RODRIGUES*

Instituto de Ciências Matemáticas de São Carlos
Universidade de São Paulo, Sao Paulo, Brazil

I. Introduction

In this chapter we shall compare the solutions of the linear system

$$\dot{y} = A(t)y \tag{1}$$

with the solutions of the perturbed linear system:

$$\dot{x} = A(t)x + f(t, x) \tag{2}$$

where $x \in X$, with $X = R^n$ or $X = C^n$, $A(t)$ is an $n \times n$ matrix continuous on $J = [t_0, \infty)$, $t_0 \geq 0$, and $f(t, x)$ is an n-dimensional column vector function continuous on $J \times X$.

We will use $\|\cdot\|$ to denote any convenient norm on X.

Let $\mu \geq 0$ be a fixed integer. We pose the following problems:

Problem 1. If $y(t) \neq 0$ is a solution of (1), does there exist a family of solutions $x(t)$ of (2) such that:

$$\lim_{t \to \infty} t^\mu \|x(t) - y(t)\| / \|y(t)\| = 0? \tag{a}$$

On how many parameters does the family of solutions of (2) depend, under the above conditions?

Problem 2. If $x(t)$ is a solution of (2), $x(t) \neq 0$ for all sufficiently large t, does there exist a family of solutions $y(t)$ of (1) such that (a) is true?

On how many parameters does the family of solutions of (1) depend?

---

* Present address: Division of Applied Mathematics, Brown University, Providence, Rhode Island.

In this work we generalize the results of Onuchic [2] and Szmydt [3]. Our work is strongly dependent on a theorem obtained by Hartman and Onuchic [1, Theorem I.1].

II. Preliminary Lemmas

Lemma 1. Let $A(t) = (a_{ij}(t))$ be an $n \times n$ continuous matrix on J, where the following conditions are satisfied:

(a) $a_{ij}(t) = 0$ if $i < j$;
(b) $a_{ii}(t) = \alpha + \lambda(t)$, with $\alpha = $ const and $\int_{t_0}^{t} R(\lambda(s)) \, ds$ bounded on J;
(c) $a_{ij}(t)$ is bounded on J for $i \neq j$;
(d) For $m = 2, \ldots, n$,

$$\lim_{t \to \infty} \left| \int_{t_0}^{t} a_{m,\,m-1}(t_{m-2}) \, dt_{m-2} \int_{t_0}^{t_{m-2}} a_{m-1,\,m-2}(t_{m-3}) \cdots \right.$$
$$\left. \times \int_{t_0}^{t_1} a_{2,\,1}(s) \, ds \right| \Big/ t^{m-1} > 0.$$

Then for each solution $y(t) \neq 0$ of $\dot{y} = A(t)y$ there is an integer l, $0 \leq l \leq n - 1$, such that:

$$0 < \varliminf_{t \to \infty} \frac{\|y(t)\|}{t^l e^{R(\alpha)t}} \leq \varlimsup_{t \to \infty} \frac{\|y(t)\|}{t^l e^{R(\alpha)t}} < \infty$$

Lemma 2. Let $A(t) = \text{diag}(A_1(t), \ldots, A_N(t))$, where $A_j(t)$, $j = 1, \ldots, N$, are $n_j \times n_j$ matrices satisfying the hypotheses of Lemma 1. Let us denote by α_j the constant element in the diagonal of $A_j(t)$. Let $R(\alpha_1) \geq \cdots \geq R(\alpha_N)$. Then for each solution $y(t) \neq 0$ of $\dot{y} = A(t)y$, there are integers i, l, $1 \leq i \leq N$, $0 \leq l \leq n_i - 1$, such that

$$0 < \varliminf_{t \to \infty} \frac{\|y(t)\|}{t^l e^{R(\alpha_i)t}} \leq \varlimsup_{t \to \infty} \frac{\|y(t)\|}{t^l e^{R(\alpha_i)t}} < \infty$$

For a proof of the above lemmas, see [2].

Theorem 1. We suppose that $A(t)$ satisfies the hypotheses of Lemma 3. Let $f(t, x)$ be continuous and $\|f(t, x)\| \leq h(t)\|x\|$ for all $t \in J$ and $x \in X$, where $h(t)$ is a continuous function satisfying $\int_{t_0}^{\infty} h(t)t^{s+\mu} \, dt < \infty$, where $s + 1$ is the largest dimension of the blocks of $A(t)$ and $\mu \geq 0$ is a fixed integer.

Let $y(t) \neq 0$ be a solution of (1) satisfying

$$\overline{\lim_{t \to \infty}} \|y(t)\|/t^l e^{R(\alpha_q)t} < \infty$$

Then there exists a p parameter family of solutions $x(t)$ of (2) such that

$$\lim_{t \to \infty} \|x(t) - y(t)\|/t^l e^{R(\alpha_q)t} = 0$$

Remark 1. The proof of the above theorem is close to the proof contained in [2, Theorem 1], replacing the space L_0^∞ by the space of the measurable functions g taking J into X, such that

$$\lim_{t \to \infty} t^\mu g(t) = 0$$

Remark 2. If $p = 0$, then we still use the expression "p parameter family of solutions" to mean that there is at least one solution.

Remark 3. The number p of parameters is specified in the proof and it is dependent on each solution $y(t)$ of (1).

The following theorem gives a positive answer to Problem 1.

Theorem 2. We suppose the hypotheses of Theorem 1 are satisfied and we let $y(t) \neq 0$ be a solution of (1). Then there exists a p parameter family of solutions $x(t)$ of (2) such that

$$\lim_{t \to \infty} t^\mu \|x(t) - y(t)\|/\|y(t)\| = 0$$

This theorem is a consequence of Lemma 2 and Theorem 1.

Lemma 3. We suppose the hypotheses of Theorem 1 are satisfied with $f(t, x)$ linear on x. Let $U(t) = (y_1(t), \ldots, y_n(t))$ be the fundamental matrix of (1) with $U(t_0) = I$, where I is the identity matrix. Then there is a matrix $V(t) = (x_1(t), \ldots, x_n(t))$ of solutions of (2) satisfying

$$\lim_{t \to \infty} t^\mu \|x_i(t) - y_i(t)\|/\|y_i(t)\| = 0 \tag{3}$$

for $i = 1, \ldots, n$, and $V(t)$ is a fundamental matrix of (2).

This Lemma is an obvious consequence of [2, Corollary 1].

Lemma 4. We suppose the assumptions of Lemma 3 are satisfied and we let $x(t)$ be a solution of (2), $x(t) \neq 0$ for all $t \in J$. Then there is a solution $y(t)$ of (1) such that

$$\lim_{t \to \infty} t^{\mu} \|x(t) - y(t)\| / \|y(t)\| = 0$$

Proof. Let $V(t)$ be the matrix of Lemma 3. Then there are constants $c_1, \ldots, c_n$ such that

$$x(t) = c_1 x_1(t) + \cdots + c_n x_n(t)$$

We take

$$y(t) = c_1 y_1(t) + \cdots + c_n y_n(t)$$

This solution of (1) satisfies our thesis.

The following theorem gives a positive answer to Problem 2.

Theorem 3. We suppose the hypotheses of Theorem 1 are satisfied and we let $x(t)$ be a solution of (2), $x(t) \neq 0$ for all $t \in J$. Then there exists a p parameter family of solutions $y(t)$ of (1) such that

$$\lim_{t \to \infty} t^{\mu} \|x(t) - y(t)\| / \|x(t)\| = 0$$

We consider the linear system

$$\dot{\psi} = A(t)\psi + \frac{x^*(t) \cdot \psi}{x^*(t) \cdot x(t)} f(t, x(t)) = A(t)\psi + g(t, \psi) \tag{4}$$

where $x^*(t)$ is the complex conjugate of $x(t)$ and $g(t, \psi)$ is a linear function of ψ.

System (4) satisfies the hypotheses of Lemma 4 and $x(t)$ is a solution of (4).

From the above theorem there is at least one solution $y(t)$ of (1) such that

$$\lim_{t \to \infty} t^{\mu} \|x(t) - y(t)\| / \|y(t)\| = 0 \tag{5}$$

It is easy to show that there is a p parameter family of solutions $z(t)$ of (1) satisfying

$$\lim_{t \to \infty} t^{\mu} \|y(t) - z(t)\| / \|y(t)\| = 0 \tag{6}$$

From (5) and (6) it follows that

$$\lim_{t \to \infty} t^{\mu} \|x(t) - z(t)\| / \|z(t)\| = 0$$

III. Applications

Definition. Let $A(t)$ and $B(t)$ be $n \times n$ matrices, where $t \in J$. We say that $A(t)$ is t-similar ($\sim$) to $B(t)$ if there is an $n \times n$ $S(t)$ matrix such that

(a) $\dot{S}(t)$ is continuous on J;

(b) $S(t)$ and $S^{-1}(t)$ are bounded on J;

(c) $\dot{S}(t) + S(t)A(t) - B(t)S(t) = 0$ for all $t \in J$.

Remark. If $A(t) \sim B(t)$, the change of variables $y = S(t)x$ leads the system $\dot{x} = A(t)x$ to the system $\dot{y} = B(t)y$.

Definition. $A(t)$ is said to be reducible if there is a constant matrix C such that $A(t) \sim C$.

Theorem 4. Let $A(t)$ be an $n \times n$ reducible matrix, that is, $A(t) \sim C$. Let J' be the Jordan canonical form of C and $s + 1$ the largest dimension of the blocks of J'. Let $f(t, x)$ be a continuous function on $J \times X$ such that

$$\|f(t, x)\| \leq h(t)\|x\|$$

where $h(t)$ is continuous and

$$\int_{t_0}^{\infty} h(t)t^{s+\mu} \, dt < \infty$$

with $\mu \geq 0$ a fixed integer number. Then

(i) For each solution $y(t) \neq 0$ of (1) there exists a p parameter family of solutions such that

$$\lim_{t \to \infty} t^{\mu}\|x(t) - y(t)\|/\|y(t)\| = 0 \tag{7}$$

(ii) If $x(t)$ is a solution of (2), $x(t) \neq 0$ for all $t \in J$, there is a q parameter family of solutions $y(t)$ of (1) satisfying (7).

Corollary 1. Let $A(t)$ be an ω-periodic matrix with $\omega > 0$ and suppose that $f(t, x)$ satisfies the hypotheses of the above theorem. Then the same conclusions of Theorem 4 are satisfied.

REFERENCES

[1] Hartman, P., and Onuchic. N., On the asymptotic integration of ordinary differential equations, *Pacific J. Math.* **13** (1963), 1193–1207.
[2] Onuchic, N., Asymptotic relationships at infinity between the solutions of two systems of ordinary differential equations, *J. Differential Equations* **3** (1967), 47–58.
[3] Szmydt, Z. Sur l'allure asymptotique de intégrales de certains systémes d'equations différentielles non linéaires, *Ann. Polon. Math. I* **2** (1955), 253–276.

Partial Peeling

OKAN GUREL
International Business Machines Corporation
White Plains, New York

Introduction

The concept of peeling as discussed in [1] has been known since Poincaré's time. The recent interest in the field of bifurcations, the name originally coined by Poincaré, focuses on restricted application of the concept either theoretically or practically. The well-known elements discussed are singular points and limit cycles as special solutions of a given dynamic system. It was later pointed out [1, 2] that one can expect new objects constructed via peeling.

Decomposition combined with peeling leads to new mathematical concepts. An example resulting from decomposition with peeling is constructed and named homoclinic limit cycle [3]. In short, D-peeling, decomposition with peeling, unravels yet another possibility resulting from peeling, which can be properly named *partial peeling*. This peeling is unknown in two-dimensional examples. Partial peeling is an example of D-peeling such that peeling is not complete along all the manifolds surrounding the singular point undergoing peeling. Here we would like to discuss briefly an example of partial peeling in a six-dimensional space. The example is constructed for a certain chemical reaction scheme with six first-order differential equations representing its kinetics [4].

Partial Peeling via D-Peeling: An Example

The example given in [4] consists of the following equations:

$$\dot{x}_1 = mx_1 - (6x_2 + x_3)x_1/(0.05 + x_1) + 7,$$

$$\dot{x}_2 = 6x_1 - 7x_2 + 3(x_5 - x_2),$$

$$\dot{x}_3 = 0.03(x_1 - x_3),$$

$$\dot{x}_4 = 0.03(x_6 - x_4),$$

$$\dot{x}_5 = 6x_6 - 7x_5 + 3(x_2 - x_5),$$

$$\dot{x}_6 = mx_6 - (6x_5 + x_4)x_6/(0.05 + x_6) + 7.$$

The only varying parameter m can be shown to be a peeling parameter. There are two distinct possibilities:

(1) $m < 0$, there is one stable singular point at x^0.

(2) $m > 0$, the singular point at x^0 becomes a saddle point and a stable limit cycle is created around this saddle point.

By setting the transformation of variables as $y = x - x^0$ we can reduce the equations to $\dot{y} = f(y)$. The characteristic values of the characteristic equation for $y < \varepsilon$ can be obtained from the linearized equations. For example, for $m = -3.03$, the characteristic values are

$$\lambda_1 = \lambda_2 = -3.03, \qquad \lambda_3 = \lambda_4 = -3.03, \qquad \lambda_5 = -7, \qquad \lambda_6 = -13.$$

All λ's are negative, and thus x^0 is a stable node located at

$$x_1^{\ 0} = x_3^{\ 0} = x_4^{\ 0} = x_6^{\ 0} = 0.79462, \qquad x_2^{\ 0} = x_5^{\ 0} = 0.68111.$$

Similarly for $m = 3$, the characteristic values of the characteristic equation of $\dot{y} = f(y)$ are found to be

$$\lambda_1 = \lambda_2 = -0.03, \qquad \lambda_3 = \lambda_4 = +3, \qquad \lambda_5 = -7, \qquad \lambda_6 = -13.$$

This indicates that the singular point is a saddle point with both stable and unstable manifolds corresponding to negative and positive characteristic values. The saddle point is located at

$$x_1^{\ 0} = x_3^{\ 0} = x_4^{\ 0} = x_6^{\ 0} = 2.3228, \qquad x_2^{\ 0} = x_5^{\ 0} = 1.9910.$$

In this case, however, there is a stable limit cycle encircling the saddle point.

The 15 projection planes can be studied. Here, only $x_1 - x_2$, $x_1 - x_3$, and $x_1 - x_5$ are shown. On the $x_1 - x_3$ plane the stable limit cycle can easily be seen. On the other hand, the $x_1 - x_5$ projection is more startling (Figs. 1–3).

The important aspect is that trajectories can be found to approach the limit cycle as they leave from a neighborhood of the singular point (Fig. 4).

The existence of the stable and unstable manifolds of the singular point is proved by the characteristic values having negative and positive real values. If P_i is an initial point of a trajectory reaching the singular point and P_{i+1}

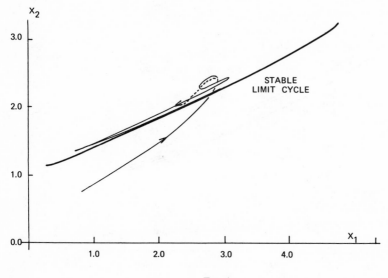

FIG. 1

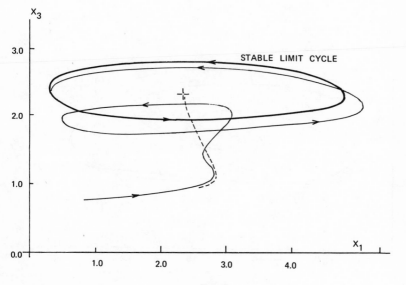

FIG. 2

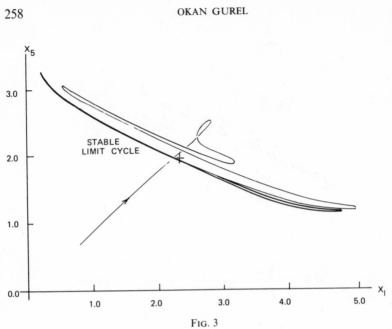

Fig. 3

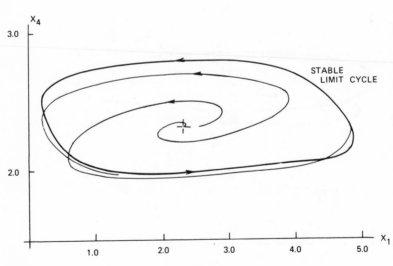

Fig. 4

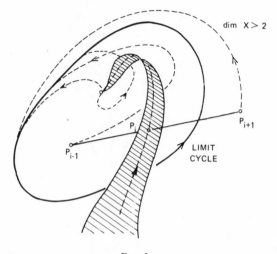

FIG. 5

another initial point for a trajectory reaching the limit cycle (Fig. 5), it can easily be detected that P_i is the intersection point of the line segment $P_{i-1} - P_{i+1}$ with the stable manifold.

REFERENCES

[1] Gurel, O., Peeling and nestling of a striated singular point, *Notices Amer. Math. Soc.* **20** (1973), A-380 and *Collective Phenomena* **2** (1975, in press).
[2] Gurel, O., Bifurcations in nerve membrane dynamics, *Internat. J. Neurosci.* **5**, (1973), 281–286.
[3] Gurel, O., Homoclinic limit cycles, *Notices Amer. Math. Soc.* **21** (1974) A-406, 407.
[4] Rössler, O. E., A synthetic approach to exotic kinetics, *in* "Physics and Mathematics of the Nervous System" (M. Conrad, W. Güttinger, and M. Dalcin, eds.), pp. 546–582. Lecture Notes in Biomathematics, Springer-Verlag, Berlin and New York, 1974.

Boundary Value Problems for Perturbed Differential Equations

T. G. PROCTOR

Department of Mathematical Sciences
Clemson University, Clemson, South Carolina

1. Introduction

Consider the question of establishing the existence of a solution of the boundary value problem

$$\dot{y} = F(t, y), \qquad My(a) + Ny(b) = V, \tag{1}$$

where F is a function from $[a, b] \times R^n$ into R^n, M and N are $n \times n$ matrices, and V is in R^n. Suppose further that F admits a decomposition into the sum of two functions

$$F(t, y) = f(t, y, \psi(t)) + g(t, y),$$

where f is such that the solutions $\phi(t, \tau, \gamma, \lambda)$ of the initial value problem

$$\dot{x} = f(t, x, \lambda), \qquad x(\tau) = \gamma, \tag{2}$$

are known for each appropriate value of the parameter λ in R^m, γ in R^n, and $a \leq \tau \leq t \leq b$, and where the functions $\psi(t, y)$, $g(t, y)$ map $[a, b] \times R^n$ into R^m and R^n, respectively. Under appropriate conditions on f, ψ, g it can be established that a function $y(t)$ is a solution of

$$\dot{y} = f(t, y, \psi(t)) + g(t, y), \qquad y(0) = \gamma, \tag{3}$$

if and only if $y(t)$ satisfies an integral equation of the form

$$y(t) = \Phi(t, \gamma) + \int_0^t H(t, s, y(s))\, ds, \tag{4}$$

where the functions Φ, H are constructed from the known functions ϕ, g, and ψ. The prototype of such an integral equation is the variation of constants formula. We will give examples later of typical hypotheses on f, ψ, and g and the resulting functions Φ, H.

From this representation for the solutions of (3), by imposing particular requirements on the boundary conditions and further requirements on ψ, g

we may now follow a familar path to establish the existence of a solution of (1). Of course, the case $M = -N = I$, $V = 0$, and the functions f, ψ, g defined and periodic for all t leads to the existence of periodic solutions of (1).

2. Results

We will first place requirements on the functions Φ, H in (4) that are sufficient to furnish the existence of a solution $y(t)$ of (4) such that

$$My(a) + Ny(b) = V.$$

Then we will give hypotheses on f, ψ, g that will guarantee that these requirements on Φ, H are satisfied.

Hypothesis H$_1$.   Let Φ be in $C([a, b] \times R^n, R^n)$ and satisfy

$$\Phi(a, \gamma) = \gamma, \qquad |\Phi(t, \gamma)| \le K|\gamma|,$$

for some $K > 0$, $a \le t \le b$, and γ in R^n. For any α in R^n suppose the equation

$$M\gamma + N(\Phi(b, \gamma) + \alpha) = V$$

has a unique continuous solution $\gamma = F(\alpha)$ and $|F(\alpha)| \le L|\alpha| + P$. For $a \le s \le t \le b$ and y in R^n, let $H(t, s, y)$ be continuous into R^n and suppose there are continuous functions $\eta_1(t, s)$, $\eta_2(t, s)$, $\ldots$, $\eta_d(t, s)$ for $a \le s \le t \le b$ and $\sigma_1(r)$, $\sigma_2(r)$, $\ldots$, $\sigma_d(r)$ for $r \ge 0$ and numbers $\theta_1, \theta_2, \ldots, \theta_d$ such that

$$|H(t, s, y)| \le \sum_{i=1}^{d} \eta_i(t, s)\sigma_i(|y|),$$

$$\overline{\lim_{r \to \infty}} \, \sigma_i(r)/r \le \theta_i, \qquad i = 1, 2, \ldots, d.$$

For $\rho > 0$ let

$$S_\rho = \{y \text{ in } C([a, b], R^n) : My(a) + Ny(b) = V, \, \|y\| \le \rho\},$$

where $\|\cdot\|$ is the uniform norm; for y in S_ρ, define

$$\alpha(y) = \int_a^b H(b, s, y(s)) \, ds,$$

$$Ty(t) = \Phi(t, F(\alpha(y))) + \int_a^t H(t, s, y(s)) \, ds,$$

for $a \leq t \leq b$; and for $i = 1, 2, \ldots, d$ let

$$Q_i = KL \int_a^b \eta_i(b, s) \, ds + \sup_{a \leq t \leq b} \int_a^t \eta_i(t, s) \, ds.$$

Theorem 1. If condition H_1 is valid and $\sum_{i=1}^d Q_i d_i < 1$ there is a $\rho > 0$ such that T has a fixed point y in S_ρ.

The proof of this theorem uses Schauder's fixed-point theorem and is adapted from the work of Fennell and Waltman [1]. This result and the other related material is discussed in detail in [4].

We will now give some conditions of f, ψ, g that lead to an integral equation of the type (4) and hypotheses on these functions that will imply H_1 is satisfied.

Hypothesis H$_2$. Let $f(t, x, \lambda)$ and the matrices of derivatives f_x and f_y be continuous for (t, x, λ) in $[a, b] \times R^{n+m}$ with $f(t, 0, \lambda) = 0$ for (t, λ) in $[a, b] \times R^m$, let $\psi(t)$ be in $C^1([a, b], R^m)$, and for λ in R^m, $a \leq s \leq b$, γ in R^n, assume that the solution $\phi(t, s, \gamma, \lambda)$ of (2) exists for $s \leq t \leq b$.

Theorem 2. Let γ be in R^n and let H_2 be satisfied. A function $y(t)$, $a \leq t \leq b$, satisfies (3) if and only if $y(t)$ satisfies (4) with

$$\Phi(t, \gamma) = \phi(t, a, \gamma, \psi(a)),$$

$$H(t, s, y) = \phi_\lambda(t, s, y, \psi(s))\dot\psi(s) + \phi_y(t, s, y, \psi(s))g(s, y).$$

The proof of this theorem appears in [3].

Hypothesis H$_3$. For $a \leq s \leq t \leq b$, $r > 0$, y in R_n, and λ in R^m, there are continuous functions $\eta_1(t, s)$, $\eta_2(t, s)$, and $\sigma(r)$ so that

$$|\phi_\lambda(t, s, y)| \leq \eta_1(t, s)|y|, \qquad |\phi_y(t, s, y, \lambda)| \leq \eta_2(t, s),$$

$$|g(t, y)| \leq \sigma(|y|), \qquad \lim_{r \to \infty} \sigma(r)/r = 0.$$

Theorem 3. Let H_2 and H_3 be satisfied and assume that for any α in R^n the equation

$$M\gamma + N(\phi(b, a, \gamma, \psi(a)) + \alpha) = V$$

has a unique solution $\gamma = F(\alpha)$, where $|F(\alpha)| \leq L|\alpha| + P$ for $P \geq 0$, $L > 0$. Then if $|\dot\psi(s)|$, $a \leq s \leq b$, is sufficiently small there is a solution of (1).

The proof of this result is given in [4]. The hypotheses were chosen to include the case $f(t, x, \lambda) = A(t, \lambda)x$, where $A(t, \lambda)$ is an $n \times n$ matrix.

Other conditions on ψ, f, and g also give that a function $y(t)$ is a solution of (3) if and only if $y(t)$ satisfies an equation of the form (3). For example, if f has continuous first and second derivatives in t, x, and λ and g has continuous derivatives in t and g, integration by parts in the equation of Theorem 2 gives

$$y(t) = \phi(t, a, \gamma, \psi(a)) + (t - a)\phi_\gamma(t, a, \gamma, \psi(a))g(a, \gamma) + \int_a^t H^*(t, s, y(s))\, ds.$$

Another case occurs when $\psi(t)$ is piecewise continuous with a piecewise continuous derivative $\dot{\psi}(t)$ except at a finite number of values of t. Finally, ψ may depend on y as well as t and still another form of (3) occurs.

It is thought that these ideas carry over to functional differential equations using results derived by Gustafson [2].

REFERENCES

[1] R. Fennell and R. Waltman, A boundary value problem for a system of nonlinear functional differential equations, *J. Math. Anal. Appl.* **26** (1969), 447–453.

[2] G. B. Gustafson, An integral equation for perturbed nonlinear functional differential equations with applications to periodic solutions and nonlinear boundary value problems, Univ. of Utah Rep. (1972).

[3] T. G. Proctor, Perio lic solutions for perturbed differential equations, *J. Math. Anal. Appl.* **47** (1974), 310–323.

[4] T. G. Proctor, Two-point boundary-value problem for perturbed differential equations, *Tech. Rep.* **174**, Dep. of Math. Sci., Clemson Univ., Clemson, South Carolina (1974).

On Stability of Solutions of Perturbed Differential Equations

B. S. LALLI* and R. S. RAMBALLY
Department of Mathematics
University of Saskatchewan, Saskatoon, Saskatchewan, Canada

1

We study the systems of differential equations of the form:

$$x' = f(t, x), \tag{1}$$

$$y' = f(t, y) + D(t)g(t, y), \tag{2}$$

where x, y, f, and g are n-vectors and $D(t)$ is a continuous $n \times n$ matrix. We assume that f and g are continuous from $I \times R^n$ to R^n, where I is the interval $[0, \infty)$. We also assume that f is continuously differentiable on $I \times D$, where D is a region in the n-dimensional x-space and that $f(t, 0) \equiv 0$. We denote by $x(t, t_0, x_0)$ the solution of (1) passing through (t_0, x_0). Similarly, let $y(t, t_0, y_0)$ be the solution of (2) through (t_0, y_0). Let $|\cdot|$ denote a convenient vector norm and a corresponding matrix norm.

It is known that $x(t, t_0, x_0)$ is a differentiable function of (t_0, x_0) on $I \times I \times D$ and that the matrix

$$\Phi(t, t_0, x_0) = \frac{\partial}{\partial x_0}\left(x(t, t_0, x_0)\right) \tag{3}$$

is the fundamental matrix of the variational system

$$z' = f_x(t, x(t, t_0, x_0))z, \tag{4}$$

which is the identity matrix for $t = t_0$.
  Moreover,

$$\frac{\partial}{\partial t_0}\left(x(t, t_0 x_0)\right) = -\Phi(t, t_0, x_0)f(t_0, x_0). \tag{5}$$

---

* The work of the first author was supported by a grant from the National Research Council of Canada and by a special grant from the President's Research Fund.

We make use of the nonlinear variation of parameters formula, which can be written

$$y(t, t_0, y_0) = x(t, t_0, y_0) + \int_{t_0}^{t} \Phi(t, s, y(s, t_0, y_0))D(s)g(s, y(s, t_0, y_0)) \, ds.$$

(6)

The following definitions will be used in the sequel. The zero solution of (1) is said to be

(a) *uniformly stable in variation* if for each $\alpha > 0$ there exists $M(\alpha)$ such that the fundamental matrix $\Phi(t, t_0, x_0)$ of (4) satisfies

$$|\Phi(t, t_0, x_0)| \le M(\alpha), \qquad t \ge t_0, \quad |x_0| \le \alpha;$$

(b) *asymptotically stable in variation* if the fundamental matrix $\Phi(t, t_0, 0)$ of $z' = f_x(t, 0)z$ satisfies

$$\int_{t_0}^{t} |\Phi(t, s, 0)| \, ds \le K, \qquad K > 0, \quad t \ge t_0.$$

(7)

The system (1) is said to be

(c) *convergent* if $\lim_{t \to \infty} x(t, t_0, x_0) = \lambda(t_0, x_0)$ is defined for each $(t_0, x_0) \in I \times R^n$;

(d) *equiconvergent* if it is convergent and for each $\varepsilon > 0$, $\alpha > 0$, $t_0 \ge 0$, there exists a function $T = T(t_0, \alpha, \varepsilon)$ such that

$$|x(t, t_0, x_0) - \lambda(t_0, x_0)| < \varepsilon, \qquad t > t_0 + T, \quad |x_0| \le \alpha;$$

(e) *equi-uniformly convergent* if the T in the above definition is independent of t_0;

(f) *equi-uniformly convergent in variation* if for each $\varepsilon > 0$, $\alpha > 0$ there exists a scalar function $T = T(\alpha, \varepsilon)$ and a matrix function $L = L(t_0, x_0)$ that is continuous on $I \times R^n$ and bounded on I such that

$$|\Phi(t, t_0, x_0) - L(t_0, x_0)| < \varepsilon, \qquad t > t_0 + T, \quad |x_0| \le \alpha.$$

2

For the first theorem we make the following assumption:

(H) There exists a continuous function $\omega(t, r)$ from $I \times I$ to I that is monotone, nondecreasing in r for each fixed t such that the scalar equation

$$r' = \omega(t, r), \qquad r(t_0) = k, \quad k > 0,$$

(8)

has a unique solution. Furthermore, the solutions of (8) exist for $t \geq t_0$ and $\lim_{t \to \infty} r(t) = r_\infty < \infty$.

The proof of Theorem 2.1 is the same as that of [4, Theorem 1].

Theorem 2.1. Assume that (H) holds and that the zero solution of (1) is uniformly stable in variation. Also suppose that

$$|\Phi(t, s, \gamma)D(t)g(t, \gamma)| \leq \omega(s, \gamma), \qquad t, s \in I, \quad \text{any } \gamma. \qquad (9)$$

Then given $\alpha > 0$ and any γ, $|\gamma| \leq \alpha$, there is a solution $y(t, t_0, \gamma) = y(t)$ of (2); for each such solution there is a solution $\bar{x}(t)$ such that

$$\lim_{t \to \infty} (y(t) - \bar{x}(t)) = 0. \qquad (10)$$

Theorem 2.2. Assume (H) and let the zero solution of (1) be uniformly stable in variation. Assume that

$$|D(t)g(t, \gamma)| \leq \omega(t, \gamma), \qquad \gamma > 0.$$

Then for every $\varepsilon > 0$ there exist $T = T(\varepsilon)$ and $\delta = \delta(\varepsilon) > 0$ such that

$$|y(t, t_0, y_0) - x(t, t_0, y_0)| < \varepsilon$$

for $t \geq t_0 \geq T$, $|y_0| < \delta$.

Theorem 2.3. Assume that the zero solution of (1) is asymptotically stable in variation. Let

$$|D(t)g(t, y)| \leq \omega(t, |y|), \qquad |y| \leq \alpha,$$

where $\omega(t, r)$ is a scalar, continuous, monotone, nondecreasing function in r for each fixed t such that

$$\int_0^\infty \omega(s, \alpha) \, ds \leq \alpha/2(\alpha L + R).$$

Then if $|y_0|$ is sufficiently small, $|y(t, t_0, y_0)| \leq \alpha$ for $t \geq t_0$. Here α is as in [2, Theorem 1].)

The proofs of Theorems 2.4 and 2.5 are the same as those of Theorems 1 and 2, respectively, of [7].

Theorem 2.4. Suppose that the system (1) is convergent and equi-uniformly convergent in variation. Assume that for each $\alpha > 0$,

$$|D(t)g(t, y)| \leq \omega_\alpha(t, |y|), \qquad |y| \leq \alpha,$$

where $\omega_\alpha(t, r)$ is a scalar, continuous, monotone, nondecreasing function in r for each fixed t such that

$$\int_0^\infty \omega_\alpha(s, c)\, ds < \infty, \qquad 0 \le c \le \alpha.$$

Then for any initial position y_0 there exists a $T = T(y_0)$ such that if $t_0 \ge T$, then the solution $y(t, t_0, y_0)$ is convergent.

Let y_0 be fixed and let $T_0(y_0)$ be the infimum of all $T(y_0)$ for which Theorem 2.4 holds. Let

$$\bar{D} = \{(t, y)\,|\,y \in R^n, t > T_0(y_0)\}.$$

Theorem 2.5. Assume that the system (1) is equiconvergent and equi-uniformly convergent in variation. Let the condition on $D(t)g(t, y)$ be as in Theorem 2.4. Then the system (2) is equiconvergent on $\bar{D}$.

3

We now study the quasilinear systems

$$x' = A(t, D(t)x)x, \tag{11}$$

$$y' = A(t, D(t)y)y + B(t, D(t)y), \tag{12}$$

where $D(t)$, $t \ge 0$, is a bounded, continuous, nonsingular matrix and where A and B are defined and continuous from $I \times R^n$ to R^n. We give sufficient conditions for the solutions of (12) to converge to any prescribed vector $\gamma \in R^n$.

Let C^l denote the space of n-vectors of functions $g(t)$ that are defined on I and such that $\lim_{t \to \infty} g(t)$ exists. For $f \in C^l$, define

$$|f|_b = \sup_{t \in I} |f(t)|,$$

where $|\cdot|$ is the Euclidean norm. With this norm C^l is a Banach space. Let

$$C_\gamma^l = \left\{ f \in C^l \,\middle|\, \lim_{t \to \infty} f(t) = \gamma \right\}.$$

We have the following theorem.

Theorem 3.1.   Assume that the matrix A satisfies

$$|A(t, D(t)u)| \leq \omega(t, |D(t)u|),$$

where $u(t) \in C_\gamma{}^l$ and where $\omega(t, r)$ is a scalar, continuous, monotone, nondecreasing function in r for each fixed t such that

$$\int_0^\infty \omega(t, c)\, dt < \infty, \qquad 0 < c < \infty.$$

Assume also that

$$\liminf_{n \to \infty} \frac{1}{n} \int_0^\infty \sup_{|u| \leq n} |B(t, D(t)u)|\, dt = 0.$$

Then there exists a solution of (12) in $C_\gamma{}^l$.

Remark.   The proof of this theorem is an application of Schauder's fixed-point theorem to an appropriate subset of $C_\gamma{}^l$. The line of reasoning to be followed is the same as in the proof of the main theorem of [6].

REFERENCES

[1] C. Avramescu, Sur l'existence des solutions convergentes d'équations différentielles non lineaires, *Ann. Mat. Pura Appl.* **4** (1969), 147–168.

[2] F. Brauer, Perturbations of nonlinear systems of differential equations, IV, *J. Math. Anal. Appl.* **37** (1972), 214–222.

[3] F. Brauer and A. Strauss, Perturbations of nonlinear systems of differential equations, III, *J. Math. Anal. Appl.* **31** (1970), 37–48.

[4] R. E. Fennel and T. G. Proctor, On asymptotic behavior of perturbed nonlinear systems, *Proc. Amer. Math. Soc.* **31** (1972), 499–504.

[5] P. Hartman, "Ordinary Differential Equations." Wiley, New York, 1964.

[6] A. G. Kartsatos and G. J. Michaelides, Existence of convergent solutions to quasilinear systems and asymptotic equivalence, *J. Differential Equations* **13** (1973), 481–489.

[7] B. S. Lalli and R. S. Rambally, Convergence of solutions of perturbed nonlinear differential equations, *Rend. Del. Cl. Sci. Fis. Mat. Nat.* **52** (1972), 850–855.

Stability Theory for Nonautonomous Systems

JOHN R. HADDOCK*
Department of Mathematical Sciences
Memphis State University, Memphis, Tennessee

We consider the nonautonomous system of differential equations

$$x' = f(t, x), \tag{1}$$

where $f: [0, \infty) \times R^n \to R^n$ is continuous. The purpose of this chapter is to employ Liapunov theory to determine behavior of solutions of (1) relative to some closed set $H \subseteq R^n$. In general, no invariance principle will be readily available and the main goal here is to present results that improve two of the classical Liapunov theorems. In particular, we give an improvement of the LaSalle–Yoshizawa theorem for nonautonomous systems for solutions approaching a set (cf. [1, p. 60]) and an improvement of the Marachkoff asymptotic stability theorem (cf. [2, p. 34]). It will be evident from Theorems 3 and 4 that we rely heavily on a close relationship between the derivative of a Liapunov function for (1) and the norm of the right-hand side of (1). In addition to presenting improved versions of the two classical theorems mentioned above, we wish to discuss briefly various other directions this author as well as several other authors have taken utilizing $V'_{(1)}$ related to $|f|$. We use the same notation as in [1], and for definitions and other terminology, we refer to [2]. We assume the reader has these two references available.

The LaSalle–Yoshizawa theorem and Marachkoff theorem have played such a fundamental role in influencing the work in this paper that we state them below as Theorems 1 and 2, respectively.

Theorem 1. Suppose there exist a Liapunov function $V: [0, \infty) \times R^n \to R^+$ for (1) and a continuous function $W: R^n \to R^+$ for (1) such that

(i) $|f(t, x)|$ is bounded on sets $[0, \infty) \times K$, where $K \subseteq R^n$ is compact, and

(ii) $V'_{(1)}(t, x) \le -W(x)$ for $t \ge 0$ and $x \in R^n$.

---

*This work was supported in part by a Memphis State University Faculty Research Grant.

If $x(t)$ is any solution of (1) with right-hand interval of definition $[t_0, \omega)$, $t_0 < \omega \leq \infty$, then $x(t) \to E_\infty$ as $t \to \omega-$, where $E = \{x \in R^n : W(x) = 0\}$ and $E_\infty = E \cup \{\infty\}$.

Theorem 2. Suppose $f(t, 0) \equiv 0$ and suppose there exists a Liapunov function V for (1) such that

(i) $|f(t, x)|$ is bounded on $[0, \infty) \times G$, where G is some bounded open subset of R^n containing the origin,

(ii) V is positive definite, and

(iii) $V'_{(1)}$ is negative definite.

Then the set $\{0\}$ (or zero solution) is asymptotically stable.

Theorems 1 and 2 are strikingly similar in that they both require $|f(t, x)|$ to be bounded in some sense and $V'_{(1)}$ to be negative definite with respect to the "attracting set." The remainder of this paper is mostly dedicated to explaining that these two restrictions can be removed while obtaining essentially the same conclusions.

Theorem 3. Suppose $H \subseteq R^n$ is closed and suppose for each $\varepsilon > 0$ and each compact $K \subseteq R^n$ there exist $\delta = \delta(\varepsilon, K) > 0$ and $\pi = \pi(\varepsilon, K) \geq 0$ such that

$$V'_{(1)}(t, x) \leq -\delta |f(t, x)| + e(t), \tag{2}$$

for $t \geq \pi$ and $x \in K \cap S^c(H, \varepsilon)$, where $\int^\infty |e(t)|\, dt < \infty$ and $S^c(H, \varepsilon)$ is the complement of the ε-neighborhood of H. If $x(t)$ is any solution of (1), then either

$$x(t) \to p \qquad \text{as} \quad t \to \infty \tag{3}$$

for some constant $p \notin H$ or

$$x(t) \to H_\infty = H \cup \{\infty\} \qquad \text{as} \quad t \to \omega-. \tag{4}$$

Theorem 4. In Theorem 3, suppose $V(t, 0) \equiv 0$, $H = \{0\}$, and that there exists $\alpha > 0$ with the property that for each $\varepsilon > 0$ $(\varepsilon < \alpha)$ there exists $\delta = \delta(\varepsilon) > 0$ and $\pi = \pi(\varepsilon) \geq 0$ such that (2) holds for $t \geq \pi$ and $\varepsilon \leq |x| \leq \alpha$. Then $H = \{0\}$ is eventually stable and any solution of (1) that remains in the "region of stability" tends to a constant as $t \to \infty$.

Remark 1. Theorem 3 is a generalization of the main theorem of [3] and is perhaps the way this result should have originally been stated.

The second theorem in [3] can then be used to show that Theorem 1 here is a special case of Theorem 3 even for $e(t) \equiv 0$. Furthermore, Eq. (2) of [3] provides an example for which the conditions of Theorem 3 hold while the conditions of Theorem 1 do not. In fact, in this example, $|f(t, x)|$ is not necessarily bounded and V' is not necessarily negative definite with respect to the set being examined. Likewise, unlike the autonomous case, an example in [4, p. 279] can be used to see that solutions in Theorem 3 do not approach the largest invariant subset of H. Analogous comments are also appropriate regarding Theorems 2 and 4 of this paper as related to Theorems 2.1 and 2.2 of [5]. Finally, if one is only concerned with bounded solutions, then $e(t)$ in (2) can be replaced by $e(t, x)$, where $\int^\infty |e(t, \bar{x}(t))|\, dt < \infty$ for bounded, continuous $\bar{x}(t)$.

Remark 2.   In Theorem 3, one would like to be able to conclude that all bounded solutions tend to H, or in Theorem 4, that all solutions in the stability region tend to zero. That is, one would like to show that (3) cannot hold. However, simple examples exist illustrating that additional minor restrictions are required in order to obtain these stronger conclusions (see, for example [3, Eq. (2)] and [5, p. 396]). In [6–8] these extra conditions are imposed on $|f|$ to ensure that solutions tend to H (or $\{0\}$). In [9–10] an uncountable family of Liapunov functions is used to eliminate a property such as (3) in Theorem 3. From [3, 5, and 8], it can be seen that a closer examination of V' is often all that is necessary to avoid (3). But even if none of these additional assumptions hold, Theorem 3 provides fairly complete knowledge of solutions that do not approach H and these solutions have not traditionally been considered.

Remark 3.   We now conclude this paper, as promised, by briefly mentioning other directions that have been taken relating V' to $|f|$. Burton [6] was perhaps the first to consider this type of condition and he proved sufficient conditions for a set H (and not H_∞) to be a global attractor. Corne and Rouche [9] also established similar conditions for a set to be a global attractor without assuming the differential equation to be defined on this set. As mentioned in Remark 2, they also employed a (possibly uncountable) family of Liapunov functions in their theory. Recently, Salvadori [10] used ideas similar to those of [9] and gave a generalization of Marachkoff's theorem that is different from our Theorem 4. Grimmer and Haddock [8] proved stability properties of certain sets using an uncountable family of Liapunov functions different from [9] and [10]. Also, some of LaSalle's results in [1] for autonomous systems are extended in [8]. Erhart [7] has taken a different direction by proving equiboundedness and perturbation theorems. Instability results using V' and $|f|$ have been established by Brown [11]. Finally, in [12] properties of V and V' have been combined to extend and improve classical boundedness results. With the aid of an invariance principle, improved boundedness results for autonomous systems are also given.

ACKNOWLEDGMENTS

The author is deeply indebted to James W. Hatley for many fruitful discussions and to Eric N. Travis for interesting comments concerning this work.

REFERENCES

[1] J. P. LaSalle, Stability theory for ordinary differential equations, *J. Differential Equations* **4** (1968), 57–65.

[2] T. Yoshizawa, Stability theory by Liapunov's second method, Math. Soc. of Japan (1966).

[3] J. R. Haddock, On Liapunov functions for nonautonomous systems, *J. Math. Anal. Appl.* **47** (1974), 599–603.

[4] J. P. LaSalle, An invariance principle in the theory of stability, in "Differential Equations and Dynamical Systems," pp. 277–286. Academic Press, New York, 1967.

[5] J. R. Haddock, Some refinements of asymptotic stability theory, *Ann. Mat. Pura Appl.* **89** (1971), 393–401.

[6] T. A. Burton, Some extensions of Liapunov's direct method, *J. Math. Anal. Appl.* **28** (1969), 545–552; **32** (1970), 689–691.

[7] J. V. Erhart, Lyapunov theory and perturbations of differential equations, *SIAM J. Math. Anal.* **4** (1973), 417–432.

[8] R. C. Grimmer and J. R. Haddock, Stability of bounded and unbounded sets for ordinary differential equations, *Ann. Mat. Pura Appl.* **99** (1974), 143–153.

[9] J. Corne and N. Rouche, Attractivity of closed sets proved by using a family of Liapunov functions, *J. Differential Equations* **13** (1973), 231–246.

[10] L. Salvadori, Some contributions to asymptotic stability theory, *Ann. Soc. Sci. Bruxelles* **88** (1974), 183–194.

[11] J. M. Brown, Some Results on Instability, Ph.D. Thesis, Southern Illinois Univ. (1971).

[12] J. R. Haddock, On Liapunov functions and boundedness and global existence of solutions, *Appl. Anal.* **2** (1973), 321–330.

A Nonoscillation Result for a Forced Second-Order Nonlinear Differential Equation

JOHN R. GRAEF* and PAUL W. SPIKES*

Department of Mathematics
Mississippi State University, Mississippi State, Mississippi

1. Introduction

In 1955 Atkinson [1] published what has become a classic work on the oscillatory behavior of solutions of second-order nonlinear differential equations. Since that time many such results have been obtained for equations of the type

$$(a(t)x')' + q(t)f(x) = r(t), \qquad (*)$$

especially when $r(t) \equiv 0$. Even though there are many results that guarantee that the above equation is oscillatory, there are relatively few that guarantee that it possesses a nonoscillatory solution (see [2, 3] for references). Even fewer sufficient conditions for nonoscillation are known, and in most cases they are for equations less general than $(*)$ (again see [2, 3]). In fact, except for some results on linear equations [6, 7], the authors know of no such results when $r(t) \not\equiv 0$ except for those in [2–5]. Complete details of the results in this paper will appear in [2].

2. A Nonoscillation Theorem

We consider the equation

$$(a(t)x')' + q(t)f(x) = r(t), \qquad (1)$$

where $a, q, r: [t_0, \infty) \to R$ and $f: R \to R$ are continuous, $a(t) > 0$, $q(t) > 0$, and $xf(x) > 0$ if $x \neq 0$. Let $q'(t)_+ = \max\{q'(t), 0\}$ and $q'(t)_- =$

---

*Supported by the Mississippi State University Biological and Physical Sciences Research Institute.

max $\{-q'(t), 0\}$ so that we have $q'(t) = q'(t)_+ - q'(t)_-$. A similar decomposition will be used for $a'(t)$. We make the following assumptions on Eq. (1):

$$f'(x) \geq 0 \qquad \text{for all} \quad x, \tag{2}$$

$$\int_{t_0}^{\infty} [q'(s)_+/q(s)]\, ds < \infty, \tag{3}$$

$$a(t) \leq a_2, \tag{4}$$

$$\int_{t_0}^{\infty} [a'(s)_-/a(s)]\, ds < \infty, \tag{5}$$

$$r(t) \geq 0, \tag{6}$$

$$\int_{t_0}^{\infty} r(s)\, ds < \infty. \tag{7}$$

Theorem. Suppose conditions (2)–(7) hold and $f'(x)$ is nondecreasing for $x \geq 0$. If for every $k > 0$,

$$\int_{t_0}^{\infty} q(s) f(ks)\, ds < \infty,$$

then all solutions of (1) are nonoscillatory.

Remark. By reversing the inequality in (6) and requiring that $f'(x)$ be nonincreasing for $x \leq 0$,

$$\int_{t_0}^{\infty} r(s)\, ds > -\infty \quad \text{and} \quad \int_{t_0}^{\infty} q(s) f(-ks)\, ds > -\infty \qquad \text{for every} \quad k > 0,$$

we can again obtain that all solutions of (1) are nonoscillatory.

We will outline a proof of this theorem in the next section, but first we wish to call the reader's attention to the conditions on $r(t)$. The equation

$$x'' + x/t^3 = [t \sin (\ln t) - 3t \cos (\ln t) + \sin (\ln t)]/t^4, \qquad t \geq 1,$$

satisfies all the hypotheses of our theorem except (6). Here $r(t)$ actually changes signs and $x(t) = [\sin (\ln t)]/t$ is an oscillatory solution of this equation. On the other hand, the equation

$$x'' + x^3/t^5 = r(t), \qquad t \geq 30,$$

where

$$r(t) = e^t + 2e^t \cos t + 3e^t(1 + \sin t)/t^7 + e^{3t}(1 + \sin t)^3/t^5$$
$$- 3e^{2t}(1 + \sin t)^2/t^6 - 2/t^3 - 1/t^8$$

satisfies all the hypotheses of the theorem except (7). This equation possesses the oscillatory solution $x(t) = e^t(1 + \sin t) - 1/t$. This last example is somewhat surprising in view of the fact that condition (7) is not required when $f(x) = x$ (see [6, 7]).

3. Proof of the Theorem

The proof of the above theorem makes use of the following three lemmas. Proofs of the lemmas will appear in [2].

Lemma 1. Suppose $r(t) \equiv 0$, conditions (2)–(4) hold, $a(t) \geq a_1$, and $k > 0$ is given. If $\int_{t_0}^{\infty} q(s)f(ks)\, ds < \infty$, then there exists a solution $x(t)$ of (1) and $T \geq t_0$ such that $x(t) \geq a_1 kt/4a_2$ for $t \geq T$.

Lemma 2. Suppose conditions (3) and (5) hold, $\int_{t_0}^{\infty} |r(s)|\, ds < \infty$, and $x(t)$ is a solution of (1). Then there exists $A > 0$ and $T \geq t_0$ such that $|x(t)| \leq At$ for $t \geq T$.

Lemma 3. Suppose $r(t) \equiv 0$, conditions (2) and (4) hold, $f'(x)$ is non-decreasing for $x \geq 0$, and $x(t)$ is a solution of (1) such that $x(t) > 0$ for $t \geq b \geq t_0$. If $u(t)$ is a function of class C' such that $u(b) = 0$ and $0 \leq u(t) \leq x(t)$ on $[b, c]$, then

$$\int_b^c q(t)f^2(u(t))\, dt \leq \int_b^c a(t)f'(u(t))[u'(t)]^2\, dt.$$

In order to prove the theorem suppose that (1) has a solution $u(t)$ that is either oscillatory (actually changes signs) or Z-type (has arbitrarily large zeros but is ultimately nonnegative or nonpositive). It is relatively easy to show that if $r(t) \geq 0$ $[r(t) \leq 0]$, then Eq. (1) can have no non-positive [nonnegative] Z-type solutions. By Lemma 2, there exist $A > 0$ and $T \geq t_0$ such that $|u(t)| \leq At$ for $t \geq T$. Lemma 1 guarantees the existence of a solution $x(t)$ of the unforced equation and a number $T_1 \geq T$ such that $x(t) \geq 2At$ for $t \geq T_1$. Now if $u(t)$ is a Z-type solution, then $u(t)$ is ultimately nonnegative, so in either case there are consecutive zeros b and c of $u(t)$ such that $T_1 < b < c$ and $u(t) > 0$ for $b < t < c$.

Notice that

$$[a(t)u'(t)f(u(t))]' = f(u(t))(a(t)u'(t))' + a(t)f'(u(t))[u'(t)]^2$$
$$= f(u(t))(-q(t)f(u(t)) + r(t)) + a(t)f'(u(t))[u'(t)]^2$$
$$\geq -q(t)f^2(u(t)) + a(t)f'(u(t))[u'(t)]^2.$$

Integrating, we have

$$a(t)u'(t)f(u(t)) \geq \int_b^t \{a(s)f'(u(s))[u'(s)]^2 - q(s)f^2(u(s))\} \, ds \geq 0,$$

by Lemma 3. This implies that $u'(t) \geq 0$ for $b \leq t \leq c$, which is a contradiction.

REFERENCES

[1] F. V. Atkinson, On second-order nonlinear oscillations, *Pacific J. Math.* **5** (1955), 643–647.

[2] J. R. Graef and P. W. Spikes, A nonoscillation result for second order ordinary differential equations, *Rend. Accad. Sci. Fis. Mat. Napoli* **41** (4) (1974), 3–12.

[3] J. R. Graef and P. W. Spikes, Nonoscillation theorems for forced second order nonlinear differential equations, *Atti Acad. Naz. Lincei Rend. Cl. Sci. Fis. Mat. Natur.* (to appear).

[4] J. R. Graef and P. W. Spikes, Sufficient conditions for nonoscillation of a second order nonlinear differential equation, *Proc. Amer. Math. Soc.* **50** (1975), 289–292.

[5] J. R. Graef and P. W. Spikes, Sufficient conditions for the equation $(a(t)x')' + h(t, x, x') + q(t)f(x, x') = e(t, x, x')$ to be nonoscillatory, *Funkcial. Ekvac.* **18** (1975), 35–40.

[6] M. S. Keener, On the solutions of certain linear nonhomogeneous second-order differential equations, *Appl. Anal.* **1** (1971), 57–63.

[7] M. Švec, On various properties of the solutions of third- and fourth-order linear differential equations, *in* "Differential Equations and Their Applications," *Proc. Conf.*, *Prague, September 1962*, pp. 187–198. Academic Press, New York, 1963.

Convexity Properties and Bounds for a Class of Linear Autonomous Mechanical Systems*

R. H. PLAUT†

Lefschetz Center for Dynamical Systems
Division of Applied Mathematics
Brown University, Providence, Rhode Island

Introduction

The vibrations and stability of the motion $r(t)$ governed by the system of n linear equations

$$(U - \eta_1 E_1 - \eta_2 E_2)r + G\dot{r} + \ddot{r} = 0 \tag{1}$$

will be considered. Here U, E_j, and G are real constant $n \times n$ matrices with U symmetric positive-definite and G skew-symmetric, while η_1 and η_2 are independent real parameters. Letting $r(t) = qe^{i\omega t}$, one obtains

$$Cq \equiv [U - \eta_1 E_1 - \eta_2 E_2 + i\omega G - \omega^2 I]q = 0, \tag{2}$$

and the corresponding characteristic equation

$$\Delta(\eta_1, \eta_2, \omega^2) \equiv \det |U - \eta_1 E_1 - \eta_2 E_2 + i\omega G - \omega^2 I| = 0. \tag{3}$$

Properties of the characteristic surfaces in the $\omega^2 - \eta_1 - \eta_2$ space will be reviewed, with special emphasis on convexity and bounds for various cases.

Case A. $G = 0$, E_j Symmetric

The *fundamental* characteristic surface is the one closest to the origin (see Fig. 1a, b, which depicts characteristic curves in the $\omega^2 - \eta_j$ and $\eta_1 - \eta_2$ planes, respectively).

---

* This research was supported by the National Science Foundation under Grant GK-41223 and the Army Research Office, Durham, under Grant DA-ARO-D-31-124-73-G130.

† Present address: Department of Civil Engineering, Virginia Polytechnic Institute and State University, Blacksburg, Virginia.

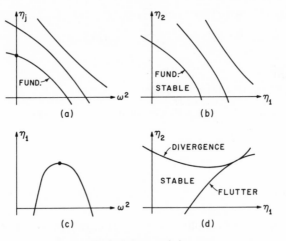

FIG. 1. Typical characteristic curves.

Theorem 1. The fundamental characteristic surface is concave toward the origin [1–3].

This result is useful for determining bounds on the fundamental surface. In the region of positive ω^2, η_1, and η_2, for example, the plane containing the intercepts of the fundamental surface with the three axes furnishes a lower bound, while the tangent planes at these three points provide an upper bound. (It is assumed in Cases A, B, and D that the fundamental characteristic surface intersects the positive η_1 and η_2 axes.)

Theorem 2. If a ray emanating from the origin intersects all the characteristic surfaces, then (i) the final surface is convex toward the origin, and (ii) all tangent planes to the characteristic surfaces intersect this ray between the fundamental and final surfaces [4, 5].

The distance from the origin to the intersection of a ray with one of these tangent planes can be written as a Rayleigh quotient, and Theorems 1 and 2 then follow from the extremum properties of this quotient.

Case B. $G = 0$, $U - \eta_1 E_1 - \eta_2 E_2$ **Symmetrizable**

Assume that $U - \eta_1 E_1 - \eta_2 E_2$ is similar to a symmetric matrix via a positive-definite transformation matrix, for any values of η_1 and η_2, so that the roots ω^2 of (3) are real.

Theorem 3.   Theorems 1 and 2 are valid for Case B [5, 6].

If the matrices E_1 and E_2 are replaced by their symmetric parts, the resulting system is called the *corresponding conservative system*.

Theorem 4.   The fundamental characteristic surface of the corresponding conservative system provides a lower bound for that of the given system [5, 6].

Case C. $G = 0$, E_j **Not Necessarily Symmetric**

In this case the onset of instability may occur by *divergence* when a root ω^2 of (3) is zero (Fig. 1a) or by *flutter* when two roots coalesce and subsequently become complex (Fig. 1c).

Theorem 5.   At a flutter point,

$$p'q = 0, \tag{4}$$

where q and p are the corresponding right and left eigenvectors of C, respectively [7].

Theorem 6.   At a flutter point, the sum of the minors of the diagonal elements of C is equal to zero [8].

Theorem 7.   If $n = 2$, then the portions of the characteristic curves (divergence instability) bounding the region of stability in the $\eta_1 - \eta_2$ plane are convex toward the region of stability (Fig. 1d) [7].

The locus of points in the $\eta_1 - \eta_2$ plane corresponding to flutter instability comprises the *flutter boundary*.

Theorem 8.   If $n = 2$, then the flutter boundary is convex toward the region of real ω^2, and in particular toward the region of stability (Fig. 1d) [7].

Theorem 8 can be proved using the extremum properties of the generalized Rayleigh quotient

$$R(x, y) = y'Ux/y'Ex, \tag{5}$$

where x and y are orthogonal n-vectors, $E = l_1 E_1 + l_2 E_2$, and l_j are the direction cosines of a ray emanating from the origin in the $\eta_1 - \eta_2$ plane [4]. For $n > 2$ the flutter boundary may be convex in some sections and concave in others [9].

Case D. $G \neq 0$, E_j Symmetric

If $G = 0$, the resulting system is called the *corresponding nongyroscopic system*.

Theorem 9. In the region of positive ω^2, η_1, and η_2, the fundamental characteristic surface of the corresponding nongyroscopic system provides an upper bound for that of the given system [10].

In the $\eta_1 - \eta_2$ plane, the fundamental characteristic curve (divergence instability) is concave toward the region of stability, as in Case A; convexity properties of the flutter boundary, however, can only be proved for special cases (e.g., see [11], where rotating shafts are considered and $G = \sqrt{\eta_2} F$).

REFERENCES

[1] P. F. Papkovich, "Works on the Structural Mechanics of Ships," Vol. 4, Moscow, 1963 (in Russian).

[2] J. D. Renton, Buckling of frames composed of thin-wall members, *in* "Thin-Walled Structures" (A. H. Chilver, ed.), pp. 1–59. Wiley, New York, 1967.

[3] K. Huseyin and J. Roorda, The loading-frequency relationship in multiple eigenvalue problems, *J. Appl. Mech.* **38** (1971), 1007–1011.

[4] K. Huseyin and R. H. Plaut, Extremum properties of the generalized Rayleigh quotient associated with flutter instability, *Quart. Appl. Math.* **32** (1974), 189–201.

[5] K. Huseyin and R. H. Plaut, Application of the Rayleigh quotient to eigenvalue problems of pseudoconservative systems, *J. Sound Vibrat.* **33** (1974), 201–210.

[6] K. Huseyin and H. Leipholz, Divergence instability of multiple-parameter circulatory systems, *Quart. Appl. Math.* **31** (1973), 185–197.

[7] K. Huseyin and R. H. Plaut, The elastic stability of two-parameter nonconservative systems, *J. Appl. Mech.* **40** (1973), 175–180.

[8] R. H. Plaut and K. Huseyin, A direct method of stability analysis for elastic circulatory systems, *in* "Developments in Mechanics" (*Proc. Midwestern Mech. Conf. 13th*), Vol. 7, pp. 425–435. Univ. of Pittsburgh, Pennsylvania, 1973.

[9] W. Hauger, On the convexity of the stability boundary of continuous elastic systems under two-parameter nonconservative loading, *Mech. Res. Commun.* **1** (1974), 107–112.

[10] K. Huseyin and R. H. Plaut, Transverse vibrations and stability of systems with gyroscopic forces, *J. Struct. Mech.* **3** (1974), 163–177.
[11] K. Huseyin and R. H. Plaut, Divergence and flutter boundaries of systems under combined conservative and gyroscopic forces, *in Dynam. Rotors, Proc. IUTAM Symp. Lyngby, Denmark, August 12-16, 1974.*

Dynamical Systems Arising from Electrical Networks

T. MATSUMOTO
Department of Electrical Engineering
Waseda University, Shinjuku, Tokyo, Japan

1. Introduction

Brayton and Moser [1] and Brayton [2] discussed several geometric aspects of electrical networks. Smale [3] observed that the dynamics of a class of electrical networks can be viewed as flows on nontrivial manifolds. Desoer and Wu [4] obtained several results along this line.

Here, we will first give a formula for describing the dynamics of a fairly general class of nonlinear networks including transistors, vacuum tubes, and various electronic devices. Second, we will discuss a property that is called reciprocity of networks. Finally, we will briefly discuss forced degeneracy.

2. The Dynamics

Consider a nonlinear network containing ρ resistors, γ capacitors, λ inductors, and v independent sources. Let q, φ, v, and i be capacitor charges, inductor fluxes, branch voltages, and branch currents, respectively. We partition v and i as

$$v = (v_R, v_C, v_L, v_e, v_j) \quad \text{and} \quad i = (i_R, i_C, i_L, i_e, i_j),$$

respectively, where R, C, L, e, and j denote resistors, capacitors, inductors, independent voltage sources, and independent current sources, respectively. Let v' and v'' be the number of independent voltage sources and the number of independent current sources, respectively, and set $b = \rho + \gamma + \lambda + v$.

There are three constraints that must be satisfied by a network. Let $\mathscr{X} = R^\gamma \times R^\lambda$ and let $\mathscr{Y} = R^b \times R^b$.

Constraint 1. Branch Characteristics. $(q, \varphi, v, i) \in \Lambda$, where Λ is a $(2b - \rho)$-dimensional differentiable submanifold of $\mathscr{X} \times \mathscr{Y}$.

Note that couplings among elements of different kinds are allowed. Pick a tree for the graph defined by the given network. Let B and Q be the fundamental loop matrix and fundamental cut set matrix, respectively.

Constraint 2. Kirchhoff Laws. $Bv = 0$, $Qi = 0$.

Constraint 3. Maxwell's Equations. $dq/dt = i_C$, $d\varphi/dt = v_L$.

Lemma 1. The set $K = \{(v, i) \in \mathcal{Y} \mid Bv = 0, Qi = 0\}$ is a b-dimensional linear subspace of $\mathcal{Y}$.

K is called the Kirchhoff space. It can be shown that K does not depend on the choice of a tree.

Assumption 1. Λ and $\mathcal{X} \times K$ are transversal, i.e., at each (x, y) in $\mathcal{X} \times \mathcal{Y}$, either $(x, y) \notin \Lambda \cap (\mathcal{X} \times K)$ or $(x, y) \in \Lambda \cap (\mathcal{X} \times K)$, and

$$T_{(x, y)}(\mathcal{X} \times \mathcal{Y}) = T_{(x, y)}(\Lambda) + T_{(x, y)}(\mathcal{X} \times K).$$

Lemma 2. Under Assumption 1, the set $\Sigma = \Lambda \cap (\mathcal{X} \times K)$ is a $(\gamma + \lambda + \nu)$-dimensional submanifold.

Σ will be the state space where the dynamics takes place. Assumption 1 is satisfied for many electrical networks and we assume that this condition is always satisfied. Note, however, that this is independent of the mathematical fact that Assumption 1 is a generic property.

Consider 1-forms on $\mathcal{X} \times \mathcal{Y}$:

$$\zeta = \sum_{n=1}^{\gamma} i_{C_n} dv_{C_n} - \sum_{n=1}^{\lambda} v_{L_n} di_{L_n} + \sum_{k=1}^{\nu'} i_{e_k} dv_{e_k} - \sum_{k=1}^{\nu''} v_{j_k} di_{j_k},$$

$$\eta = \sum_{n=1}^{\rho} v_{R_n} di_{R_n} + d\left(\sum_{n=1}^{\gamma} v_{C_n} i_{C_n} + \sum_{k=1}^{\nu'} v_{e_k} i_{e_k} \right).$$

Lemma 3. Let $\iota_1^*: K \to \mathcal{X} \times \mathcal{Y}$ be inclusion. Then $\iota_1^*\zeta = \iota_1^*\eta$.

Proof. It suffices to show that

$$\sum_{k=1}^{b} v_k \, di_k = 0 \qquad \text{on} \quad K.$$

To show this let z be a (global) coordinate for K. Then there are real numbers a_{kn} and b_{kn} with

$$v_k = \sum_{n=1}^{b} a_{kn} z_n, \qquad i_k = \sum_{n=1}^{b} b_{kn} z_n, \qquad k = 1, \ldots, b.$$

By Tellegen's theorem

$$\sum_{k=1}^{b} \sum_{n=1}^{b} a_{kn} b_{kn} (z_n)^2 = 0,$$

for all z in R^b. Hence

$$\sum_{k=1}^{b} a_{kn} b_{kn} = 0, \qquad n = 1, \ldots, b,$$

so that

$$\sum_{k=1}^{b} v_k \, di_k = \sum_{k=1}^{b} \sum_{n=1}^{b} a_{kn} b_{kn} z_n \, dz_n = 0 \qquad \text{on} \quad K.$$

Using Lemma 3, one can write the dynamics: Let $\iota_2 : \Sigma \to K$ be inclusion and let $\iota = \iota_1 \circ \iota_2$. Let (W, ψ) be a chart of Σ and let $\pi_{q_n} : (x, y) \to q_n$ be the natural projection. Write

$$q_n(z) = (\pi_{q_n} \circ \iota \circ \psi^{-1})(z), \qquad z \in \psi(W).$$

Similar notations will be used for other variables. Let $\omega = \iota^* \eta$.

Theorem 1. The dynamics of the network is described by

$$\sum_{n=1}^{\gamma} \frac{dq_n(z)}{dt} dv_{C_n}(z) - \sum_{n=1}^{\lambda} \frac{d\varphi_n(z)}{dt} di_{L_n}(z) + \sum_{k=1}^{v'} i_{e_k}(z) \, dv_{e_k}(z)$$

$$- \sum_{k=1}^{v''} v_{j_k}(z) \, di_{j_k}(z) = \omega^{\#}(z), \qquad z \in \psi(W),$$

where $\omega^{\#} = (\psi^{-1})^* \omega$.

3. Reciprocity

We call a network *reciprocal* if

$$\iota^* \left(\sum_{n=1}^{\rho} dv_{R_n} \wedge di_{R_n} \right) = 0. \tag{3.1}$$

Theorem 2. Suppose that Σ is simply connected (this is true for many networks). Then the network is reciprocal if and only if there is a smooth real-valued function P on Σ with $\omega = dP$.

Proof. If (3.1) holds, then

$$d\omega = d(\imath^*\eta) = \imath^*\,d\eta$$

$$= \imath^*\left(\sum_{n=1}^{\rho} dv_{R_n} \wedge di_{R_n} + d^2\left(\sum_{n=1}^{\gamma} v_{C_n} i_{C_n}\right) + d^2\left(\sum_{k=1}^{\gamma'} v_{e_k} i_{e_k}\right)\right) = 0,$$

so that ω is a closed 1-form on Σ. Since Σ is simply connected, ω is exact on Σ. Conversely, if $\omega = dP$, then ω is exact and a fortiori closed. The result follows from a direct computation.

Reciprocity is important in that the dynamics is essentially determined by the function P. P is called the *mixed potential function* or the *generalized potential function*. Various definitions for reciprocity have been given. All of them, however, are concerned with the properties of Λ. The fact that

$$\sum_{n=1}^{\rho} dv_{R_n} \wedge di_{R_n} = 0 \qquad \text{on} \quad \Lambda \tag{3.2}$$

implies (3.1). But the converse may not be true. It is easy to give an example of a network that is reciprocal in the sense of (3.1) but does not satisfy (3.2). Theorem 2 seems to justify the naturalness of definition (3.1) in the sense that reciprocity is a necessary and sufficient condition for the existence of the generalized potential function. If one needs only sufficiency, (3.2) will suffice.

Corollary. If Λ is simply connected and if (3.2) holds, then there is a function P on Σ with $\omega = dP$.

Let $x = (q, \varphi)$, $y = (i_e, v_j)$, and $u = (v_e, i_j)$. In many electrical networks (x, u) serves as coordinate for Σ.

Theorem 3. Suppose that the network is reciprocal and that Σ is simply connected. If (x, u) serves as a coordinate of Σ, then the dynamics is described by

$$\left(\frac{dx}{dt}, y\right)\begin{pmatrix} A(x, u) & 0 \\ B(x, u) & C(x, u) \end{pmatrix} = \frac{\partial}{\partial(x, u)} P^{\#}(x, u), \tag{3.3}$$

where

$$A(x, u) = (a_{nm}(x, u)), \qquad B(x, u) = (b_{nk}(x, u)), \qquad C(x, u) = \operatorname{diag} c_k(x, u),$$

$$P^{\#} = P \circ \psi^{-1}$$

with

$$a_{nm}(x, u) = \begin{cases} \partial v_{C_n}(x, u)/\partial x_m, & n = 1, \ldots, \gamma, & m = 1, \ldots, \gamma + \lambda, \\ -\partial i_{L_n}(x, u)/\partial x_m, & n = \gamma + 1, \ldots, \gamma + \lambda, & m = 1, \ldots, \gamma + \lambda, \end{cases}$$

$$b_{nk}(x, u) = \begin{cases} \partial v_{C_n}(x, u)/\partial u_k, & n = 1, \ldots, \gamma, & k = 1, \ldots, \nu, \\ -\partial i_{L_n}(x, u)/\partial u_k, & n = \gamma + 1, \ldots, \gamma + \lambda, & k = 1, \ldots, \nu, \end{cases}$$

$$c_k(x, u) = \begin{cases} -i_{e_k}(x, u) & k = 1, \ldots, \nu', \\ v_{j_k}(x, u), & k = \nu' + 1, \ldots, \nu. \end{cases}$$

x, y, and u are sometimes called *state*, *output*, and *input*, respectively. Note that both sides of (3.3) are understood to be row vectors.

From a circuit-theoretic point of view, one would like to have a reasonably simple formula to check reciprocity. Let (W, ψ) be a chart of Σ and write

$$v_R(z) = (\pi_{v_R} \circ \iota \circ \psi^{-1})(z), \qquad i_R(z) = (\pi_{i_R} \circ \iota \circ \psi^{-1})(z), \qquad z \in \psi(W).$$

Proposition. A network is reciprocal if and only if for each chart (W, ψ) and for all z in $\psi(W)$,

$$\frac{\partial v_R}{\partial z}(z)^{\mathrm{T}} \frac{\partial i_R}{\partial z}(z) = \frac{\partial i_R}{\partial z}(z)^{\mathrm{T}} \frac{\partial v_R}{\partial z}(z),$$

where T denotes matrix transposition.

Brayton's result for n-ports [2] is a special case of this proposition.

4. Forced Degeneracy

Consider the special case

$$\Lambda = \{(q, \varphi, v, i) \in \mathscr{X} \times \mathscr{Y} \,|\, (q, v_C) \in \Lambda_C,$$

$$(\varphi, i_L) \in \Lambda_L, (v_R, i_R) \in \Lambda_R, (v_e, i_e, v_j, i_j) \in \Lambda_S\},$$

where Λ_C, Λ_L, Λ_R, and Λ_S are submanifolds of appropriate dimensions. Let t and l be the number of tree branches and the number of links, respectively. It is known that rank $B = l$, rank $Q = t$, and $t + l = b$, so that

$$\dim \{v \in R^b \,|\, Bv = 0\} = t, \qquad \dim \{i \in R^b \,|\, Qi = 0\} = l.$$

Now if $t < \gamma$, then some of the components of i_C have no constraints so that they cannot be parametrized by any other variables in the network. Similarly, if $l < \lambda$, then some of the components of v_L cannot be parametrized by any other variables in the network. Maxwell's equations then tell us that the differential equation is not well defined. Such situations are called *forced degeneracy* [3]. This degeneracy can be resolved in a certain manner. Since space is limited, we will not give detailed arguments. The basic idea is to construct a generalized Kirchhoff space that has a lower dimension than that of K.

ACKNOWLEDGMENTS

The author is deeply indebted to Prof. Y. Ishizuka, Waseda University, and Prof. C. P. Ong, University of California, for many helpful discussions.

REFERENCES

[1] R. Brayton and J. Moser, A theory of nonlinear networks, *Quart. Appl. Math.* **22** (1964), 1–33, 81–104.

[2] R. Brayton, Nonlinear reciprocal networks, *in SIAM–AMS Proc.* **III** (1971), 1–15.

[3] S. Smale, On the mathematical foundations of electrical circuit theory, *J. Differential Geometry* **7** (1972), 193–210.

[4] C. Desoer and F. Wu, Trajectories of nonlinear RLC networks, *IEEE Trans. Circuit Theory* **19** (1972), 562–570.

An Invariance Principle for Vector Liapunov Functions

JOSEPH P. LASALLE* and E. N. ONWUCHEKWA**†
Lefschetz Center for Dynamical Systems
Division of Applied Mathematics
Brown University, Providence, Rhode Island

In the definition of a Liapunov function V by LaSalle in [5] the differential equation was assumed to be defined on an open set G^* of R^n and V was taken to be locally Lipschitz continuous on an arbitrary set G with $\overline{G} \subset G^*$. Motivated by applications to economics, we want to drop the assumption that $\overline{G} \subset G^*$ and to replace local Lipschitz continuity by continuity.

Consider

$$\dot{x} = f(x), \qquad f : G^* \subset R^n \to R^n, \tag{1}$$

where G^* is an open set of R^n and f is continuous on G^*. In addition, we assume that solutions are unique. (This condition is not necessary but then the discussion is quite complex, and the uniqueness assumption is almost always acceptable in applications.) For $x^0 \in G^*$, $x(t, x^0)$ denotes the solution of (1) satisfying $x(0, x^0) = x^0$. Define

$$\dot{V}(x^0) = \liminf_{t \to 0^+} \frac{V(x(t, x^0)) - V(x^0)}{t}. \tag{2}$$

Definition 1. Let G be an arbitrary set of R^n, $G \subset G^*$. (Here it can be that $G = G^*$.) We say that V is a *Liapunov function* of (1) on G if $V(x)$ is continuous on G^* and if $\dot{V}(x) \le 0$ for all $x \in G$.

---

* This research was supported in part by the Office of Naval Research, NONR N00014-67-A-0191-000906, in part by the Air Force Office of Scientific Research, AF-AFOSR 71-2078C, in part by the National Science Foundation, GP 28931X2, and in part by the U.S. Army Research Office, Durham, DA-ARO-D-31-124-73-G-130.

** This research was supported in part by the National Science Foundation, GP 28931X2, and in part by the Government of the Federal Republic of Nigeria.

† Present address: Ministry of Economic Development, Port Harcourt, Nigeria.

Lemma 1. Let V be a Liapunov function of (1) on G and let $x^0 \in G$. Suppose that $x(t, x^0)$ is defined and $x(t, x^0) \in G$ for all $t \in [0, \beta)$ and some $\beta > 0$. Then $V(x(t, x^0))$ is differentiable almost everywhere on $[0, \beta)$, and on $[0, \beta)$

$$V(x(t, x^0)) - V(x^0) \le \int_0^t \dot{V}(x(\tau, x^0)) \, d\tau \le 0.$$

For $x^0 \in G^*$ the solution $x(t, x^0)$ has a maximum positive interval of definition $[0, \omega)$ relative to G^*. This means that $x(t, x^0) \in G^*$ for all $t \in [0, \omega)$, and if ω is finite, $x(t, x^0)$ can have no positive limit points in G^*. [p is a positive limit point of $x(t, x^0)$ if there is a sequence t_n such that $t_n \to \omega^-$ and $x(t_n) \to p$ as $n \to \infty$.]

If V is a Liapunov function of (1) on G, define $E = \{x; \; \dot{V}(x) = 0, x \in \overline{G} \cap G^*\}$, and let M be the largest invariant set in E. M is the union of all solutions defined on $(-\infty, \infty)$ that remain in E for all t.

Using Lemma 1, we obtain an invariance principle for autonomous differential equations as follows:

Theorem 1. Let V be a Liapunov function for (1) on G and $x(t)$ a solution of (1) with maximal positive interval of definition $[0, \omega)$. Suppose that $x(t)$ is in G for all t in $[0, \omega)$. Then

   (i) $x(t, x^0) \to \infty$ as $t \to \omega^-$,
  (ii) $x(t)$ has positive limit points on the boundary of G^*, or
 (iii) $\omega = \infty$ and $x(t, x^0) \to M_\infty = M \cup \{\infty\}$ as $t \to \infty$.

The proof of this theorem is similar to that given by LaSalle in [5]. The basic idea goes back to 1960 (see the paper by LaSalle in Volume 1, Chapter 5, of this treatise).

A special case motivated by a stability problem in economics was given by Uzawa [9] in 1961 (there E was the set of equilibrium points in $\overline{G} \cap G^*$, $E = M$). Economists are not aware of this generalization of Liapunov's classical theory, and it is only recently that we have seen these applications to economics. We now turn to the use of vector Liapunov functions to obtain information on the positive limit sets of solutions.

There are many instances in applications (for example, [3]) where it is advantageous to use more than one Liapunov function to study the stability and the asymptotic behavior of systems. Perhaps the simplest result is the following (also expressible by the fact that sum of a finite number of Liapunov functions is a Liapunov function), which is a direct consequence of Theorem 1: A solution $x(t)$ of (1) is said to be *compact* if $x(t)$ is contained in a compact set relative to G^* [i.e., $x(t)$ is bounded for $t \geq 0$ and has no positive limit points on the boundary of G^*].

Corollary 1. Let V_i ($i = 1, \ldots, m$) be Liapunov functions of (1) on G. If $x(t)$ is a compact solution of (1) that remains in G for all $t \geq 0$, then $x(t) \to M = \bigcap_{i=1}^{m} M_i$, where M_i is the largest invariant subset of $E_i = \{x; V_i(x) = 0, x \in \overline{G} \cap G^*\}$.

A more interesting method for constructing a Liapunov function from a number of functions was used by Arrow et al. [1], which has been applied quite often in the study of stability in economics (see, for instance, [2, 4, 9]). Suppose that $V_1, \ldots, V_m$ are continuous real-valued functions on G^*. Let $J = \{1, 2, \ldots, m\}$, and define

$$W(x) = \underset{j \in J}{\text{Max}}\ V_j(x).$$

Then it is easy to see that $W(x)$ is continuous and, if the V_j are C^1, that

$$\dot{W}(x) = \underset{j \in N(x)}{\text{Max}}\ \dot{V}_j(x),$$

where $N(x) = \{i; V_i(x) = V(x)\}$. Hence, if $\dot{V}_j(x) \leq 0$ for each $x \in G$ and $j \in N(x)$, $W(x)$ is a Liapunov function on G. This was in essence the observation made by Arrow et al. [1]. To formalize this, define $V(x) = (V_1(x), \ldots, V_m(x))$. We shall then say that $V(x)$ is a *vector Liapunov function of* (1) *on* G if (1) $V: G^* \to R^m$ is C^1, and (2) $\dot{V}_j(x) \leq 0$ for each $j \in N(x)$ and each $x \in G$. Let $E = \{x; \dot{V}_i(x) = 0$ for some $i \in N(x), x \in \overline{G} \cap G^*\}$, and define M to be the largest invariant set in E. We then have

Corollary 2.   Let V be a vector Liapunov function of (1) on G. Then every solution of (1) that is compact and remains in G for all $t \geq 0$ approaches M as $t \to \infty$.

We now give a sufficient condition that a C^1 function V mapping R^n into R^n be a vector Liapunov function. We denote the Jacobian matrix of V by V' ($V'_{ij} = \partial V_i / \partial x_j$). Then

$$\dot{V}(x) = (\dot{V}_1(x), \ldots, \dot{V}_n(x)) = V'(x) f(x).$$

Proposition 1. Suppose $V: G^* \to R^n$ is C^1. If for each $x \in G$ there exists a vector $c(x)$, $c_j(x) \neq 0$ for $j = 1, \ldots, n$, such that $V_i(x) \geq V_j(x)$, $j = 1, \ldots, n$, implies

(1^0) $V'_{ij}(x)c_j(x) \geq 0, j \neq i$,
(2^0) $(V'(x)c(x))_i \leq 0$,
(3^0) $0 \leq f_i(x)/c_i(x) \geq f_j(x)/c_j(x), j = 1, \ldots, n$,

then V is a vector Liapunov function of (1) on G.

Let us give an interpretation of the above result to show that the conditions are not as odd as they may at first seem. Suppose that

$$\dot{x} = f(x) = F(x, u(x)) \tag{3}$$

is a model for the spread of an infectious disease through population groups $P_1, \ldots, P_n$, where $x_i(t)$ is the number of infected individuals in P_i at time t. Let $G^* = R^n$ and $G = \{x; 0 \leq x_i \leq P_i, i = 1, \ldots, n\}$ (p_i is the number of individuals in P_i). By the nature of the model, G will be positively invariant. The vector $u(x) = (u_1(x), \ldots, u_m(x))$ represents a control law for the control factors $u_1, \ldots, u_m$ available for the cure and prevention of the disease. Then $V_i(x) = W_i(x, u(x))$ are relative measures of the social cost of the disease plus the control cost for the population groups P_i. Suppose it were possible to find a control law $u(x)$ for which (1^0)–(3^0) were satisfied on G and, moreover, such that $(V'(x)c(x))_i < 0$ for each $x \in G$ not an equilibrium state of (3). Then $V(x)$ is a vector Liapunov function on G and M is the set of all equilibrium states of (2) in G. The control brings the system to M (M is an attractor relative to G). For instance, if the cost of control were unimportant, one could take $V_i(x) = x_i$. The question then would be whether for $x \in G$, $x \neq 0$, there is sufficient control to make $f_i(x) < 0$ when $x_i \geq x_j, j = 1, \ldots, n$. If such a control exists, then the origin can be made globally asymptotically stable relative to G. The control $u(x)$ may be a natural feedback that exists in a society dependent on education, morals, cost of treatment, legislation, etc. An analogous interpretation for (3) as a price adjustment process in a free or controlled economy is the type of application discussed in [1, 2, 4].

As another example, suppose $G = G^* = R^n$, $f(0) = 0$, and f is C^1 on R^n. Then (1) is of the form

$$\dot{x} = A(x)x. \tag{4}$$

The matrix $A(x)$ is not unique but $A(x) = \int_0^1 f'(sx) \, ds$ is always one such matrix. Taking $V_i(x) = x_i^2$, we then see that $V(x)$ is a Liapunov function on

R^n if for $0 < x_i^2 \geq x_j^2$, $j = 1, \ldots, n$, the diagonal term of the ith row of $A(x)$ is negative dominant, i.e., $a_{ii}(x) + \sum_{j \neq i} |a_{ij}(x)| < 0$. It then follows that (4) is globally asymptotically stable. This is the case, for instance, if the diagonal of the Jacobian $f'(x)$ is negative dominant $[f'_{ii}(x) + \sum_{j \neq i} |f'_{ij}(x)| < 0$ for all $i]$ for all $x \neq 0$. This result is related to a conjecture by Arrow and Hahn (see [2, p. 295], and [4, p. 251]).

The particular notion of a vector Liapunov function that we have formalized here seems to arise naturally in many applications, and results of this type can be found in [8]. A more detailed discussion of vector Liapunov functions and applications of the type indicated here will be presented in future papers [6, 7].

REFERENCES

[1] K. J. Arrow, H. D. Block, and L. Hurwitz, On the stability of competitive equilibrium, II, *Econometrica* **27** (1959), 82–109.

[2] K. J. Arrow and F. H. Hahn, "General Competitive Analysis." Holden-Day, San Francisco, California, 1971.

[3] L. T. Grujic and D. D. Siljak, Asymptotic stability and instability of large-scale systems, *IEEE Trans. AC* **AC-18** (1973), 636–645.

[4] G. Horwich and P. A. Samuelson (eds.), "Trade, Stability, and Macroeconomics." Academic Press, New York, 1974.

[5] J. P. LaSalle, Stability theory for differential equations, *J. Differential Equations* **4** (1968), 57–65.

[6] J. P. LaSalle and E. N. Onwuchekwa, Stability by partitioning and applications (to appear).

[7] J. P. LaSalle and E. N. Onwuchekwa, Some remarks on the stability of generalized Metzlerian systems (to appear).

[8] E. N. Onwuchekwa, Stability of differential equations with applications to economics, Ph.D. dissertation, Brown Univ., Providence, Rhode Island, 1975.

[9] H. Uzawa, The stability of dynamic processes, *Econometrica* **29** (1961), 617–631.

Stability of a Nonlinear Volterra Equation

J. A. NOHEL* and D. F. SHEA**
Department of Mathematics
University of Wisconsin, Madison, Wisconsin

1. Introduction

We study the stability and asymptotics of the identically zero solution of the scalar Volterra equation

$$x'(t) = -\int_0^t g(x(\xi))a(t - \xi)\,d\xi + f(t, x(t)), \qquad ' = d/dt, \qquad (1.1)$$

on the interval $0 \le t < \infty$. Equation (1.1) arises in several applications and has been studied by Levin and Nohel [1, Theorems 2 and 3] under hypotheses that are considerably more restrictive than required here, both with respect to the positivity of $a(t)$ and the smoothness of $g(x)$ and $f(t, x)$. The present generalization is possible because of results obtained by us in [2] concerning the behavior of solutions of (1.1) when $f(t, x)$ is independent of x and in $L^1(0, \infty)$ with respect to t. Detailed proofs and further discussion of the present results may be found in [3].

The method used to study (1.1) combines energy functions (different from those used in [1]) with the frequency domain method developed in [2]. The method can also be used to study the stability properties of the Volterra equation with infinite delay on $0 \le t < \infty$:

$$x'(t) = -\int_{-\infty}^t g(x(\xi))a(t - \xi)\,d\xi + f(t, x(t)), \qquad (1.2)$$

which may be regarded as a special case of the Volterra–Stieltjes equation

$$x'(t) = -\int_0^t g(x(t - \xi))\,dA(\xi) + F(t; x), \qquad (1.3)$$

where $F(t; x)$ is a nonlinear functional on a suitable Banach space. In another direction the perturbations $f(t, x)$ in (1.1), rather than being small

* Supported by ARO Grant DAH C04-74-G0012.
** Supported by NSF Grant GP-21340.

with respect to x in the sense of Theorem 1 below, may be small because of the presence of a small parameter, say μ, and we may study the equation

$$x'(t) = -\int_0^t g(x(\xi))a(t-\xi)\,d\xi + \mu f(t, x(t)) \tag{1.4}$$

by our method as well.

2. Results and Discussion

Concerning $a(t)$, we assume

$$a(t)e^{-\sigma t} \in L^1(0, \infty) \qquad \text{for all} \quad \sigma > 0. \tag{2.1}$$

Letting $\pi = \{s \in C : \operatorname{Re} s > 0\}$ define

$$\hat{a}(s) = \int_0^\infty e^{-st}a(t)\,dt, \qquad s \in \pi,$$

$$U(i\tau) = \liminf_{s \to i\tau,\, s \in \pi} \operatorname{Re} \hat{a}(s), \qquad -\infty \le \tau \le \infty,$$

$$U_0(i\tau) = \liminf_{\sigma \to 0^+} \operatorname{Re} \hat{a}(\sigma + i\tau),$$

$$Q_a[v; T] = \int_0^T v(t) \int_0^t v(t)a(t-\tau)\,d\tau\,dt, \qquad T > 0, \quad v \in C[0, T].$$

Let $g(x)$ be continuous and define

$$G(x) = \int_0^x g(\xi)\,d\xi,$$

$$g_0(x) = \begin{cases} \max_{0 \le \xi \le x} g(\xi) & \text{if} \quad x \ge 0, \\[2mm] \min_{x \le \xi \le 0} g(\xi) & \text{if} \quad x \le 0, \end{cases}$$

$$M(x) = \max(g_0(x), -g_0(-x)), \qquad x \ge 0,$$

$$m(x) = \min(G(x), G(-x)).$$

Theorem 1. (i) Let $a(t)$ satisfy (2.1) and let $U(i\tau) \ge 0$ for $-\infty \le \tau \le \infty$. Let $g(x) \in C(-\infty, \infty)$, $xg(x) \ge 0$, and $f(t, x) \in C(0 \le t < \infty, -\infty < x < \infty)$, with $g(x)$ not identically zero in any neighborhood of the origin.

For every $\varepsilon > 0$ let there exist $\delta = \delta(\varepsilon) > 0$ and $\beta(t) = \beta(t, \varepsilon) \geq 0$ such that $\delta(\varepsilon) \to 0$ $(\varepsilon \to 0)$, $\int_0^\infty \beta(t, \varepsilon)\, dt \leq \varepsilon$, and such that

$$|f(t, x)| \leq \beta(t), \qquad 0 \leq r < \infty, \quad |x| \leq \delta, \tag{2.2}$$

$$m(\delta) > \varepsilon M(\delta), \tag{2.3}$$

for $\varepsilon > 0$ sufficiently small.

Then for any $x_1 > 0$ there exists $x_0 = x_0(x_1) > 0$ such that every solution $x(t)$ of (1.1) with $|x(0)| \leq x_0$ exists on $0 \leq t < \infty$ and $|x(t)| < x_1$ $(0 \leq t < \infty)$.

(ii) If, in addition,

$$U_0(i\tau) \geq \eta/(i + \tau^2) \qquad \text{for some} \quad \eta > 0 \text{ a.e.,} \quad -\infty < \tau < \infty, \tag{2.4}$$

$$a(t) \in BV[1, \infty), \tag{2.5}$$

then $\lim_{t \to \infty} g(x(t)) = 0$.

By a recent result of Staffens [4], it can be shown that condition (2.4) in (ii) can be replaced by the more general condition $U(i\tau) > 0$ $(-\infty < \tau < \infty)$, without changing the conclusion of Theorem 1(ii).

Examples of g, f satisfying the conditions of Theorem 1 are given in [1] and [3]; in [1] more smoothness is required than is needed here. It is proved in [2] that the condition $U(i\tau) \geq 0$ $(-\infty \leq \tau \leq \infty)$ in Theorem 1(i) is equivalent to positivity in the sense that $Q_a[v; T] \geq 0$ for every $T > 0$ and for every $v \in C[0, T]$. It is further shown in [2] that a function $a(t) \in L^1_{\text{loc}}(0, \infty)$ that is positive, nonincreasing, and convex on $(0, \infty)$ satisfies this condition, and therefore the kernels $a(t)$ considered in [1] are a very special case of this class; for other kernels $a(t)$ satisfying the hypothesis of Theorem 1(i) see [2, 3]. It is also proved in [2] that a function $a(t) \in L^1_{\text{loc}}(0, \infty)$ that is nonincreasing, positive, convex, not identically constant, and such that the measure $da'(t)$ is not purely singular satisfies condition (2.4); in particular, this includes kernels $a(t) \in L^1(0, 1)$, $(-1)^k a^{(k)}(t) \geq 0$ $(k = 0, 1, 2)$, $a(t) \neq a(0)$, and hence also kernels $a(t)$ studied in [1] as a very special case. The equivalence of condition (2.4) with strong positivity in the sense of Halanay is proved in [2]. Condition (2.5) is needed to establish the uniform continuity of the solution $x(t)$ on $[0, \infty)$, which is in turn needed to prove Theorem 1(ii).

Theorem 1(i) is proved using the energy functions

$$E(t) = G(x(t)) + Q_a[g(x(t)); t] \geq 0,$$

$$V(t) = \{\alpha^{-1} M(\delta) + E(t)\} \exp\left(-\alpha \int_0^t \beta(\tau)\, d\tau\right),$$

where $\alpha > 0$ is a suitably chosen constant (see [3]); it may be noted that these energy functions are different from those employed in [1]. It is shown in [3] using (1.1) that $V'(t) \leq 0$ and Theorem 1(i) follows from the inequality

$$0 \leq V(t) \leq V(0) = \alpha^{-1}M(\delta) + G(x(0)).$$

with an appropriate choice of $x_0 > 0$ and of the constant $\alpha > 0$, from which one shows that

$$0 \leq G(x(t)) + Q_a[g(x(t)); t] < m(\delta), \qquad 0 \leq t < \infty.$$

Theorem 1(ii) is proved from the inequality $0 \leq Q_a[g(x(t)); t] < m(\delta)$ $(0 \leq t < \infty)$ using the method (too lengthy to be given here) developed in [2].

To apply the method of Theorem 1 to the Volterra–Stieltjes functional equation (1.3), let $A \in BV[0, \infty)$ and let F be a continuous functional on $[0, \infty) \times C(-\infty, 0)$. We now let

$$U(i\tau) = U_0(i\tau) = \int_0^\infty \cos \tau t \, dA(t),$$

and we prove

Theorem 2. (i) Let $A(t)$, $F(t; x)$ be as above, with $\int_0^\infty t|dA(t)| < \infty$ and $U(i\tau) \geq 0$ $(-\infty < \tau < \infty)$. Let $g(x)$, $\delta(t)$, $\beta(t, \varepsilon)$ satisfy the assumptions of Theorem 1(i), with (2.2) replaced by

$$|F(t; \psi)| \leq \beta(t), \qquad 0 \leq t < \infty, \tag{2.6}$$

whenever $\|\psi\|_\infty = \sup_{-\infty < t \leq 0} |\psi(t)| \leq \delta$.

Then for any $x_1 > 0$ there exists $x_0 = x_0(x_1) > 0$ such that if $\varphi \in C(-\infty, 0]$ satisfies $\|\varphi\|_\infty \leq x_0$, then any local solution of (1.3) with initial function φ on $-\infty < t \leq 0$ exists on $0 \leq t < \infty$ and satisfies $|x(t)| < x_1$ $(0 \leq t < \infty)$.

(ii) If also (2.4) or $U(i\tau) > 0$ $(-\infty < \tau < \infty)$ is satisfied, then $\lim_{t \to \infty} g(x(t)) = 0$.

To apply Theorem 2 to (1.2) define

$$F(t; x) = \int_{-\infty}^0 g(\varphi(\xi))a(t - \xi) \, d\xi + f(t, x(t)),$$

where $x(t)$ is that solution of (1.2) with initial function φ on $(-\infty, 0]$; define $A(t) = \int_0^t a(\xi) \, d\xi$. If $ta(t) \in L^1(0, \infty)$, $\varphi \in C(-\infty, 0]$, $\|\varphi\|_\infty \leq x_0$, and if $f(t, x)$, $g(x)$, satisfy the hypotheses of Theorem 1, then the hypotheses of Theorem 2 are satisfied.

Another equation that may be studied by applying Theorem 2 is

$$x'(t) = -\theta g(x(t)) - \int_0^t g(x(t - \xi))a(\xi) \, d\xi + f(t, x(t)),$$

where $\theta > 0$ is a constant; to reduce this to Eq. (1.3) define $A(t) = \theta I(t) + \int_0^t a(\sigma) \, d\sigma$, where I is the unit impulse function, and take for $F(t; x)$ the function $f(t, x)$ satisfying the hypotheses of Theorem 1.

The proof of Theorem 2 (too lengthy to be given here; see [3]) uses the energy functions

$$E(t) = G(x(t)) + \tilde{Q}_A[g(x); t],$$

where

$$\tilde{Q}_A[v; t] = \int_0^t v(\tau) \int_0^\tau v(\tau - \xi) \, dA(\xi) \, d\tau$$

[it is shown that $\tilde{Q}_A[v; T] \geq 0$ for every $T > 0$ and for every $v \in C[0, T]$ if and only if $U(i\tau) \geq 0$, $-\infty < \tau < \infty$], and

$$V(t) = \{\alpha^{-1}M + E(t)\} \exp\left(-\alpha \int_0^t \left[\beta(\tau) + N \int_{(\tau, \infty)} |dA(\xi)|\right] d\tau\right),$$

where $\alpha > 0$ is a constant depending on $\varepsilon > 0$, $N = \sup_{-\infty < t \leq 0} |g(\varphi(t))|$, and where $M = M(\delta)$.

Concerning (1.4) one can prove the following result by the method of Theorem 1 (see [3]).

Theorem 3. (i) Let $a(t)$ and $g(x)$ satisfy the hypotheses of Theorem 1(i). Let $f(t, x) \in C$ $(0 \leq t < \infty, |x| < \infty)$ and let there exist $\gamma(t) \in L^1(0, \infty)$ (independent of μ) such that $|f(t, x)| \leq \gamma(t)$ $(0 \leq t < \infty, |x| < \infty)$.

Then for any $x_1 > 0$, there exist $x_0 = x_0(x_1) > 0$ and $\mu_0 = \mu_0(x_1) > 0$ such that every solution $x(t) = x(t, \mu)$ of (1.4) with $|x(0)| \leq x_0$ and $|\mu| \leq \mu_0$ exists on $0 \leq t < \infty$ and satisfies $|x(t)| < x_1$.

(ii) If in addition either (2.4) or $U(i\tau) > 0$ $(-\infty < \tau < \infty)$ and (2.5) are satisfied then $\lim_{t \to \infty} g(x(t)) = 0$.

REFERENCES

[1] Levin, J. J., and Nohel, J. A., Perturbations of a nonlinear Volterra equation, *Michigan Math. J.* **12** (1965), 431–447.

[2] Nohel, J. A. and Shea, D. F., Frequency domain methods for Volterra equations, *Advan. Math.* (to appear).

[3] Nohel, J. A. and Shea, D. F., Stability of a nonlinear Volterra equation, *Bol. Un. Mat. Ital.* (to appear).

[4] Staffens, O. J., Nonlinear Volterra integral equations with positive definite kernels, *Proc. Amer. Math. Soc.* **51** (1975), 103–108.

On a Class of Volterra Integrodifferential Equations

GEORGE SEIFERT

Department of Mathematics
Iowa State University, Ames, Iowa

The asymptotic behavior of solutions of the linear Volterra integro-differential equation

$$x'(t) = Ax(t) + \int_0^t B(t - s)x(s)\, ds, \qquad x(0) = x_0, \tag{1}$$

in case $B(t)$ is integrable on $[0, \infty)$ was studied by Miller [3] and Grossman and Miller [2]. In [2], their results were extended to the case where the derivative $B'(t)$ is integrable on $[0, \infty)$, but $B(t)$ may not be. The methods in [2] and [3] involve the use of a well-known theorem of Paley and Wiener and the resulting conditions are consequently in terms of the Laplace transforms of $B(t)$ and $B'(t)$.

More recently, systems more general than (1) have been studied using so-called Liapunov–Razumikhin conditions [1]. In particular, sufficient conditions for certain asymptotic behavior of solutions of such systems have been given where $B(t - s)$ is replaced by $B(t, s)$, to which methods in [2] and [3] based on the Paley–Wiener theorem do not apply. While the conditions given in [2] and [3] are necessary as well as sufficient for certain asymptotic behavior, the conditions in [1] are apparently not. However, since the latter do not involve Laplace transforms, but more direct conditions on A and $B(t)$, they should be of some practical interest.

We assume that in (1) $x(t)$ is a real m-vector, A and $B(t)$ are real $m \times m$ matrices, and denote by I the identity matrix. We denote by A^T the transpose of A.

As a consequence of results in [2] and [3] the following can be obtained [here and henceforth $B^*(\lambda) = \int_0^\infty e^{-\lambda t}B(t)\, dt$].

Theorem 1. Let $B(t) \in L^1(R^+)$, i.e., the matrix elements of $B(t)$ are functions integrable on $[0, \infty)$. Let

$$\det(\lambda I - A - B^*(\lambda)) \neq 0 \qquad \text{for} \quad \operatorname{Re} \lambda \geq 0. \tag{1.1}$$

Then the solution $x \equiv 0$ is asymptotically stable, i.e., stable in the usual Liapunov sense, and attracting.

Actually the hypotheses of this theorem imply a stronger stability, viz., uniform asymptotic stability (cf. [3]), which in turn, under the condition $B(t) \in L^1(R^+)$, also implies (1.1).

From the results in [1], we have the following

Theorem 2. Let $B(t) \in L^1(R^+)$,

$$\det(\lambda I - A) \neq 0 \qquad \text{for} \quad \text{Re } \lambda \geq 0, \tag{1.2}$$

and

$$2\beta \int_0^\infty |CB(t)| \, dt < \alpha, \tag{1.3}$$

where $A^\mathsf{T}C + CA = -I$, α^2 and β^2 are, respectively, the least and greatest eigenvalues of the positive definite symmetric matrix C. Then $x \equiv 0$ is asymptotically stable.

We observe that (1.2) is just the condition that A be stable.

In case $B(t) = B_1(t) + B_2(t)$ where $B_1(t) \in L^1(R^+)$, $B_2'(t) \in L^1(R^+)$, a standard trick of introducing $y(t) = \int_0^t x(s) \, ds$ and integration by parts yields the $2m$-dimensional system

$$z'(t) = \hat{A}z(t) + \int_0^t \hat{B}(t - s)z(s) \, ds, \tag{2}$$

where

$$z = \begin{pmatrix} x \\ y \end{pmatrix}, \qquad \hat{A} = \begin{pmatrix} A & B_2(0) \\ I & 0 \end{pmatrix}, \qquad \hat{B}(t) = \begin{pmatrix} B_1(t) & B_2'(t) \\ 0 & 0 \end{pmatrix}.$$

Since the asymptotic stability of $z \equiv 0$ clearly implies that of $x \equiv 0$, we have as a consequence of Theorem 1:

Theorem 3. Let $B(t) = B_1(t) + B_2(t)$, where $B_1(t)$ and $B_2(t)$ are as above. Let

$$\det(\lambda^2 I - \lambda(A + B_1{}^*(\lambda)) - B_2(0) - B_2'^*(\lambda)) \neq 0 \qquad \text{for} \quad \text{Re } \lambda \geq 0. \tag{2.1}$$

Then $x \equiv 0$ is asymptotically stable.

On the other hand, using Theorem 2, we obtain:

Theorem 4. Let $B(t)$ be as in Theorem 3. and let

$$\det(\lambda^2 I - \lambda A - B_2(0)) \neq 0 \qquad \text{for} \quad \text{Re } \lambda \geq 0, \tag{2.2}$$

and

$$2\beta \int_0^\infty |\hat{C}\hat{B}(t)|\, dt < \alpha, \tag{2.3}$$

where $\hat{A}^{\mathrm{T}}\hat{C} + \hat{C}\hat{A} = -\hat{I}$, the $2m \times 2m$ identity matrix, α^2 and β^2 are, respectively, the least and greatest eigenvalues of $\hat{C}$, and $\hat{A}$ and $\hat{B}(t)$ are as in (2). Then $x \equiv 0$ is asymptotically stable.

We again note that (2.2) is just the condition that $\hat{A}$ be stable.

As a corollary to Theorem 4 we have:

Corollary. Let $|B_1(t)| \le Me^{-\sigma t}$, $|B_2'(t)| \le Me^{-\sigma t}$ for $t \ge 0$ and constants M and σ. Then for σ sufficiently large, $x \equiv 0$ is asymptotically stable.

In case $A = -\rho I$ and $B_2(0) = -\gamma I$, $\rho > 0$, $\gamma > 0$, some fairly routine calculations yield

$$\hat{C} = \begin{pmatrix} aI & bI \\ bI & cI \end{pmatrix}$$

where

$$a = \frac{1 + \gamma}{2\rho\gamma}, \qquad b = \frac{1}{2\gamma}, \qquad c = \frac{1 + \gamma}{2\rho} + \frac{\rho}{2\gamma},$$

from which we have easily that the eigenvalues of $\hat{C}$ are $\{(a + c) \pm [(a + c)^2 + 4(b^2 - ac)]^{1/2}\}/2$. Thus condition (2.3) can be easily checked in this case.

Finally we state an extension of Theorem 4 to the nonconvolution case where $B(t - s)$ in (1) is replaced by $B(t, s)$.

Theorem 5. Let $B(t, s) = B_1(t, s) + B_2(t, s)$. Suppose $B_2(t, t) \to B_0$ as $t \to \infty$, and

(i) $\det(\lambda^2 I - \lambda A - B_0) \ne 0$ for Re $\lambda \ge 0$.

If $\hat{A}^{\mathrm{T}}\hat{C} + \hat{C}\hat{A} = -\hat{I}$, where $\hat{A} = \begin{pmatrix} A & B_0 \\ I & 0 \end{pmatrix}$ and α^2 and β^2 are, respectively, the least and greatest eigenvalues of $\hat{C}$, suppose

(ii) $2\beta \int_0^\infty |\hat{C}\hat{B}(t, s)|\, ds < \alpha \qquad$ for $t \ge 0$,

where

$$\hat{B}(t, s) = \begin{pmatrix} B_1(t, s) & B_{2s}(t, s) \\ 0 & 0 \end{pmatrix};$$

(iii) $\quad \displaystyle\lim_{T \to \infty} \sup\left\{\left|\int_0^{t-T} |B_1(t, s)|\, ds, t \geq T\right.\right\} = 0,$

and

$$\lim_{T \to \infty} \sup\left\{\left|\int_0^{t-T} |B_{2s}(t, s)|\, ds, t \geq T\right.\right\} = 0.$$

Here $B_{2s} = \partial B_2/\partial s$.

(iv) for fixed $t \geq 0$, both $\int_0^t |\hat{B}(t + h, s) - \hat{B}(t, s)|\, ds$ and $\int_0^h \hat{B}(t + h, s + h)|\, ds$ tend to 0 as $h \to 0$.

Then $x \equiv 0$ is an asymptotically stable solution of

$$x'(t) = Ax(t) + \int_0^t B(t, s)x(s)\, ds, \qquad x(0) = x_0, \tag{3}$$

The proof of Theorem 5 is based on suitably modified results in [1].

REFERENCES

[1] R. C. Grimmer and G. Seifert, Stability properties of Volterra integrodifferential equations, *J. Differential Equations* **19** (1975), 142–166.

[2] S. I. Grossman and R. K. Miller, Nonlinear Volterra integrodifferential systems with *L*-kernels, *J. Differential Equations* **13** (1973), 551–556.

[3] R. K. Miller, Asymptotic stability properties of linear Volterra integrodifferential equations, *J. Differential Equations* **10** (1971), 485–506.

Existence and Continuation Properties of Solutions of a Nonlinear Volterra Integral Equation

TERRY L. HERDMAN*

Department of Mathematics
University of Oklahoma, Norman, Oklahoma

This paper concerns questions of existence and continuation of solutions of an n-dimensional nonlinear integral equation of Volterra type,

$$x(t) = f(t) + \int_0^t g(t, s, x(s)) \, ds, \qquad t \in [0, \infty). \tag{E}$$

In particular, there are derived sufficient conditions for a solution $x(t)$ of Eq. (E) on its maximal interval of existence $[0, T)$ to possess the property that $x(t)$ tends to the boundary of an open region $\mathscr{R}$ as $t \to T^-$.

For a general $m \times n$ matrix $M = [M_{\alpha\beta}]$ the symbol $M \cdot \geq \cdot 0$ signifies that the elements of M are real, and $M_{\alpha\beta} \geq 0$ for $\alpha = 1, \ldots, m$, $\beta = 1, \ldots, n$. For $x \in \mathbf{R}^n$ the symbol $|x|$ is employed for the vector $(|x_\alpha|)$, $\alpha = 1, \ldots, n$. For $T \in (0, \infty)$ the set $\{(t, s) : 0 \leq t \leq T, 0 \leq s \leq T\}$ is denoted by $\mathbf{Q}_T$. A matrix function $M(t, s)$ is said to be an element of $\mathfrak{L}^2(\mathbf{Q}_T)$ if $M(t, s)$ is measurable and quadratically integrable in the Lebesgue sense on the square $\mathbf{Q}_T$.

The following hypotheses concerning the vector functions f and g occurring in Eq. (E) are used in the following discussion.

(H1) f is continuous on $\mathbf{R}^+$.

(H2) g is defined for all $(t, s, x) \in \mathbf{R}^+ \times \mathbf{R}^+ \times \mathbf{R}^n$, $g(t, s, x) = 0$ whenever $s > t$ and $x \in \mathbf{R}^n$; moreover, g is measurable in s on $[0, t]$ for each $(t, x) \in \mathbf{R}^+ \times \mathbf{R}^n$, and g is continuous in x for each fixed pair $(t, s) \in \mathbf{R}^+ \times \mathbf{R}^+$.

---

* Present address: Department of Mathematics, Virginia Polytechnic Institute and State University, Blacksburg, Virginia.

(H3) There exists an $n \times n$ matrix function M and an n-dimensional vector function p satisfying the following conditions:

(i) $M(t, s) \cdot \geq \cdot 0$, $p(t, s) \cdot \geq \cdot 0$ for $(t, s) \in \mathbf{R}^+ \times \mathbf{R}^+$ and $M(t, s) = 0$, $p(t, s) = 0$ whenever $s > t$.

(ii) $M \in \mathfrak{L}^2(\mathbf{Q}_T)$, $p \in \mathfrak{L}^2(\mathbf{Q}_T)$ for each $T \in (0, \infty)$.

(iii) For $(t, s, x) \in \mathbf{R}^+ \times \mathbf{R}^+ \times \mathbf{R}^n$ we have that

$$|g(t, s, x)| \cdot \leq \cdot M(t, s)|x| + p(t, s).$$

(H4) For $T \in (0, \infty)$ and B a compact set in $\mathbf{R}^n$, the function $v(t, u, x) = \int_0^u g(t, s, x(s)) \, ds$ is continuous in (t, u) on $\mathbf{Q}_T$ uniformly for $x \in \mathscr{C}([0, T]; B)$.

(H5) The functions M and p of (H3) satisfy the following conditions:

(i) There exists a $k(T; M) < \infty$ such that

$$\int_0^t \|M(t, s)\|^2 \, ds \leq k(T; M) \qquad \text{for} \quad t \in [0, T];$$

(ii) There exists a $k(T; p) < \infty$ such that

$$\int_0^t \|p(t, s)\| \, ds \leq k(T; p) \qquad \text{for} \quad t \in [0, T].$$

(H6) There is a $T' > 0$ and $\eta > 0$ for which there exists a vector function $p_0: \mathbf{Q}_T \to \mathbf{R}^n$ such that:

(i) $p_0 \in \mathfrak{L}^2(\mathbf{Q}_T)$, $p_0(t, s) \cdot \geq \cdot 0$ on $\mathbf{Q}_{T'}$ and $p_0(t, s) = 0$ whenever $s > t$;

(ii) $|g(t, s, x)| \cdot \leq \cdot p_0(t, s)$ on the set

$$\{(t, s, x) : (t, s) \in \mathbf{Q}_{T'}, \|x - f(s)\| \leq \eta\};$$

(iii) There exists a value $k(T', p_0) < \infty$ such that

$$\int_0^t \|p_0(t, s)\| \, ds < k(T', p_0) \qquad \text{for} \quad t \in [0, T'].$$

Given $T > 0$, consider Eq. (E), where the functions f and g satisfy hypotheses (H1)–(H5). The existence of a solution of (E) that is defined on the interval $[0, T]$ is established by first employing a device introduced by Tonelli [4] to provide approximate solutions, then obtaining an actual solution by a limiting process. Now there may be other solutions of (E) that exist on some subinterval, say $[0, T_1]$ or $[0, T_1)$, of the interval $[0, T]$;

Theorem A establishes the extensibility of the interval of definition of any such solution to be the whole interval $[0, T]$.

Theorem A. Suppose that hypotheses (H1)–(H5) are satisfied and that $T \in (0, \infty)$ is given. If ϕ is a solution of (E) on $[0, T_1)$ or $[0, T_1]$, where $0 < T_1 < T$, then there exists on $[0, T]$ a solution y of (E) such that $y(t) \equiv \phi(t)$ on the respective subintervals $[0, T_1)$, $[0, T_1]$.

Theorem A is established by using an argument similar to that for the differential equations case (see Reid [3, Chapter 1]).

Replacing (H3) and (H5) by (H6), we have the following local existence theorem.

Theorem B. If (H1), (H2), (H4), and (H6) are satisfied, then there exists a number $\beta > 0$ and a function y such that y is a solution of (E) on $[0, \beta]$ satisfying $\|y(t) - f(t)\| < \eta$.

By definition, an open region of the (t, s, x)-space is an open connected subset. Given an open region $\mathscr{R}$ and a solution ϕ of (E) on an interval I, with $\{(t, s, \phi(s)) : 0 \le s \le t, t \in I\}$ contained in $\mathscr{R}$, the interval I is said to be a maximal interval of existence of ϕ if there does not exist a solution $y(t)$, $t \in I_0$, of (E) with $\{(t, s, y(s)) : 0 \le s \le t, t \in I_0\}$ in $\mathscr{R}$, where $I \subset I_0$, $y(t) \equiv \phi(t)$ for $t \in I$, and there is a $t_0 \in I_0$ such that $t < t_0$ for every $t \in I$. If a half-open interval $[0, T)$ is an interval of existence for a vector function $y(t)$ with $\{(t, s, y(s)) : 0 \le s \le t, t \in [0, T)\}$ in $\mathscr{R}$, then $y(t)$ is said to tend to the boundary of $\mathscr{R}$ as $t \to T^-$ if either $T = +\infty$ or $T < +\infty$, and for S an arbitrary compact subset of $\mathscr{R}$ there is a corresponding $T_0 < T$ such that if $t \in (T_0, T)$ then $(t, t, y(t)) \notin S$.

We now state two hypotheses concerning an open region $\mathscr{R}$, which we shall employ in the next theorem.

(H7)   There exists a function $\hat{p} \colon \mathbf{R} \times \mathbf{R} \to \mathbf{R}^n$ such that

   (i) $\hat{p}(t, s) \cdot \ge \cdot 0$ and $\hat{p}(t, s) = 0$ whenever $s > t$;

   (ii)   If $T \in (0, \infty)$ then $\hat{p} \in \mathfrak{L}^2(\mathbf{Q}_T)$ and there exists a value $k(T, \hat{p}) < \infty$ such that for every $t \in [0, T]$ we have

$$\int_0^t \|\hat{p}(t, s)\| \, ds < k(T, \hat{p});$$

   (iii) $|g(t, s, x)| \cdot \le \cdot \hat{p}(t, s)$ for $(t, s, x) \in \mathscr{R}$.

(H8) For the open region $\mathscr{R}$, $T \in (0, \infty)$, and B a compact set in $\mathbf{R}^n$, the function

$$v(t, u, x) = \int_0^u g(t, s, x(s))\, ds$$

is continuous in (t, u), uniformly for $x \in \mathscr{C}([0, \ T]; \ B)$ satisfying $(t, u, x(u)) \in \mathscr{R}$ for $0 \le u \le t \le T$.

Theorem C. Suppose that $\mathscr{R}$ is an open region of (t, s, x)-space, (H1), (H2), (H7), and (H8) are satisfied, and $y_0(t)$, $t \in [0, T_0)$, is a solution of (E) with $\{(t, s, y_0(s)) : 0 \le s \le t < T_0\}$ in $\mathscr{R}$. Then there exists a solution $y(t)$, $t \in [0, T)$, of (E) that is an extension of $y_0(t)$ and such that $[0, T)$ is a maximal interval of existence for $y(t)$; moreover, $y(t)$ tends to the boundary of $\mathscr{R}$ as $t \to T^-$.

For a discussion of these topics under different hypotheses the reader is referred to Miller [2, Chapter 2]. Also, for a more detailed discussion of these results and related topics see Herdman [1].

REFERENCES

[1] T. L. Herdman, Existence and Continuation Properties of Solutions of a Non-Linear Volterra Integral Equation, Dissertation, Univ. of Oklahoma (May, 1974).
[2] R. K. Miller, "Nonlinear Volterra Integral Equations." Benjamin, New York, 1971.
[3] W. T. Reid, "Ordinary Differential Equations." Wiley, New York, 1971.
[4] L. Tonelli, Sulle equazioni funzionali del tipo de Volterra, *Bull. Calcutta Math. Soc.* **20** (1928), 31–48.

Author Index

311

Subject Index